KU-698-939

KONRAD
WACHSMANN

BUILDING THE WOODEN HOUSE

TECHNIQUE AND DESIGN

With new contributions of Christa & Michael Grüning and Christian Sumi
Translation by Peter Reuss

Birkhäuser Verlag
Basel · Boston · Berlin

First published in German 1930 under the title „Holzhausbau – Technik und Gestaltung" by Ernst Wasmuth Verlag AG, Berlin.

Library of Congress Cataloging-in-Publication Data

A CIP catalogue record for this book is available from the Library of Congress, Washington D.C., USA

Building the wooden house : technique and design / Konrad Wachsmann. Transl. by Peter Reuss. – / with new contributions of Michael Grüning and Christian Sumi. Basel ; Boston ; Berlin : Birkhäuser, 1995
Einheitssacht.: Holzhausbau <engl.>
ISBN 3-7643-5134-9
NE: Wachsmann, Konrad; Reuss, Peter [Übers.]; EST

This edition is also available in German.
Printed on acid-free paper produced of chlorine-free pulp
Cover design: Martin Schack, Dortmund
Printed in Germany
ISBN 3-7643-5134-9
ISBN 0-8176-5134-9

9 8 7 6 5 4 3 2 1

Christa & Michael Grüning

Konrad Wachsmann: Albert Einstein's Architect – Pioneer of Architectural Engineering

Konrad Wachsmann has earned a place among the leading and most original architects of this century – not only as the architect of the famous Einstein house, but as a pioneer in the field of architectural engineering.

Konrad Wachsmann's long list of friends, partners, and colleagues was inspired – as were his students – by his ideas and projects, always far in advance of their time, often utopian.

The list included many of the 20th century's leading architectural innovators: Hans Poelzig, Walter Gropius, Heinrich Tessenow, Le Corbusier, Mies van der Rohe, Buckminster Fuller, Kenzo Tange, Erich Mendelsohn, Julius Posener, Max Bill, Otto Kolb, Serge Chermayeff, Egon Eiermann, Frei Otto, Fritz Haller.

In the words of the Viennese architectural historian, Friedrich Achleitner: "Wachsmann fascinated an entire generation of architects. With young people his use of the Socratic method to pose questions, to set problems, and to respect their thoughts mobilized and inspired them to produce their best."

It is, therefore, not surprising that the works of this German-American citizen of the world influenced such differing people as Richard Rogers, Renzo Piano, Norman Foster, Arata Isozaki, and Hans Hollein. But it is surprising that now the current generation of architects and engineers has rediscovered Konrad Wachsmann and his two most influential works: *Building the Wooden House* (Holzhausbau) and *Turning Point of Building* (Wendepunkt im Bauen).

Frei Otto, whose buildings have caused sensations similar to those once raised by Wachsmann's huge hangars, explains the renaissance of Wachsmann's ideas thus: "I consider Konrad Wachsmann to be the most cogent philosopher of modern architecture. As no other, he raised technical innovation and architectural beauty to new heights."

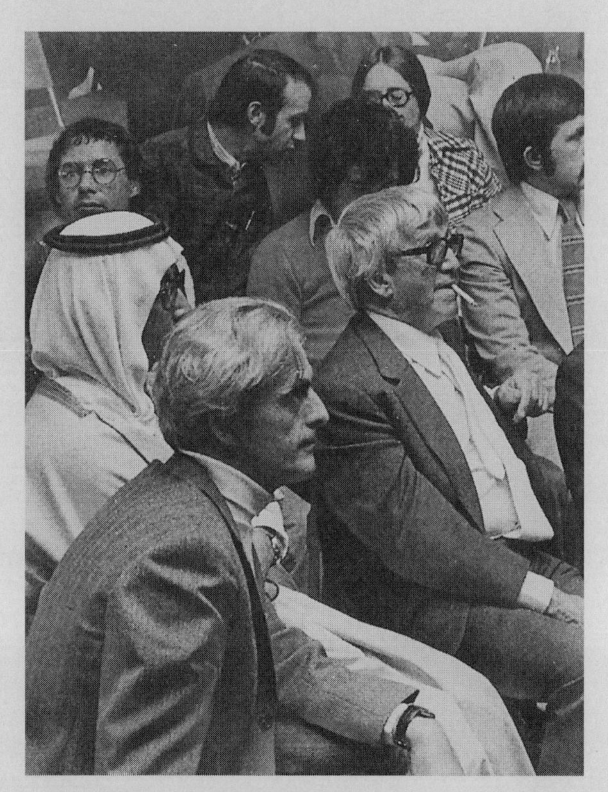

1 Frei Otto said of his friend and colleague: "For me, Konrad Wachsmann was one of the 'greats' of our profession. He was intensely involved in our research work. The University of Stuttgart awarded him an honorary degree of Doctor of Engineering, and he was elected a member of the Academy of Arts in Berlin." The photo was taken in 1978 at the Institut für leichte Flächentragwerke in Stuttgart, an institute to which Wachsmann was particularly attached.

Konrad Wachsmann was born on May 16, 1901, in Frankfurt on the Oder, into a line of well established Jewish pharmacists. In earlier times, the family had been granted privileges by the Grand Elector and, through the generations, had become fully assimilated. As Wachsmann recalled near the end of his life, "One thought of oneself then as German, lived according to the best Prussian traditions, and was a loyal – if critical – subject, first of the king, then of the emperor."

In the case of the young Wachsmann, this critical spirit, however, soon led to a break with the old conventions. After the early death of his father, the sheltered son became somewhat of a problem-child, rebelling against his bourgeois and provincial surroundings. Above all, his rebellion was directed against his schooling. As he himself related it: "I had to repeat the second and third forms. Finally the school made an arrangement with my mother: as I was again destined to repeat the pre-fourth form, the offer was made simply to give me a fourth form certificate – on the condition that my mother would voluntarily remove me from the school. My poor, unfortunate mother naturally had to agree. And that was the end of my schooling."

It appeared that this would also be the end of Wachsmann's childhood dream: to become an architect. Instead of further academic training, Wachsmann entered and completed a carpenter's and furniture-maker's apprenticeship. He built coffins for fallen soldiers and, as a seventeen-year-old, experienced the collapse of a society that he had come to despise.

But he also discovered Voltaire's *Candide*, and was greatly excited by reports of the founding of the Bauhaus in Weimar – all very disturbing for his family, that continued to be the young Wachsmann's source of financial support. Lyonel Feininger's expressionist cathedral (which decorated the Gropius Manifesto) shocked the family. (They also refused to read the Manifesto.) And so, at the family's direction, it was not in the Bauhaus in Weimar, but in the School of Arts and Crafts in Berlin that Wachsmann found himself.

2 A German family portrait. Konrad Wachsmann (2nd from left) with his sisters, Marga and Charlotte, and his brother, Heinz. Heinz died as a lieutenant in the First World War. Charlotte, her son, and Wachsmann's mother, Elsa, were killed in a concentration camp near Riga. Marga, because of her "Aryan" marriage to Wilhelm Blume, was able to escape the Nazis.

But in Berlin, the young Wachsmann, precocious and feeling himself already knowledgeable enough about furniture making, quickly discovered Berlin's night life, the theater, and the Romanische Café – which he was later to call the most important "university" of his life. There, he numbered among his acquaintances, friends, and companions many who would help broaden his horizons: Bertold Brecht; Else Lasker-Schüler and Herwarth Walden, publishers of the *Sturm*, and their circle; Tucholsky, Mehring, Ringelnatz, Grosz, Joseph Roth, Wieland Herzfelde, John Heartfield, John Höxter, Klaus Mann, Orlik, Bruno

Frei – and others. He surrounded himself with revolutionaries, Dadaists.

The dangers to which Wachsmann was being exposed did not remain long undiscovered by his family. And so, after only a year, he was forced to leave Berlin. As Wachsmann recalled when he was 78, "The School of Arts and Crafts gave me nothing; the Romanische Café, the theater, the papers, literature – everything. I even considered whether it might not make sense for me to become a writer. But without my being consulted, it was decided that I should go to Dresden. In that way I came to the city on the Elbe and became, at the Art Academy, a student of Heinrich Tessenow, one of Germany's most famous architects."

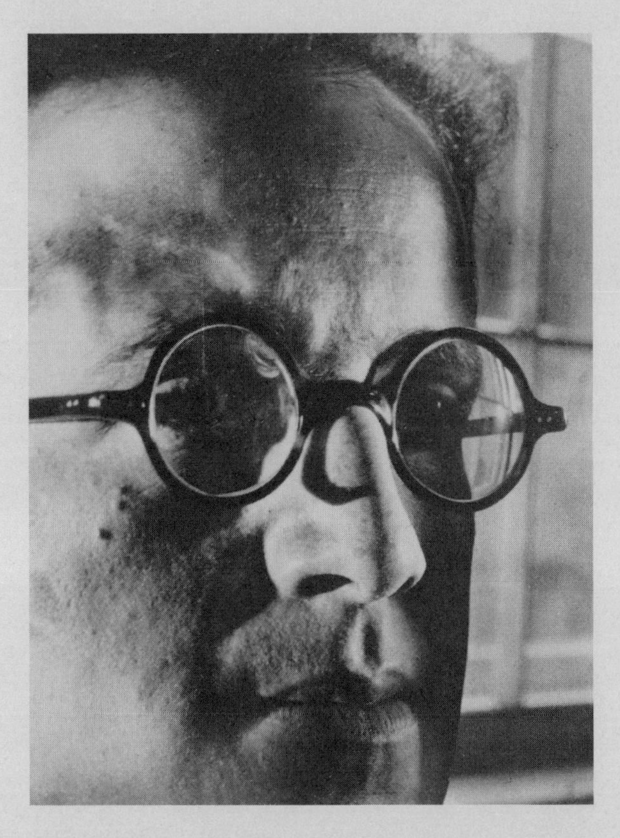

3 Konrad Wachsmann in the Romanische Café. For the young architect, the café was to be his "most important university". There he came to know the key avant-garde figures of the turbulent '20s in the fields of art, architecture, literature, and science. Photo by Laszlo Moholy-Nagy or Lucia Moholy.

Although Tessenow not only encouraged but also demanded much of his talented student, Wachsmann returned to Berlin, prematurely, in November, 1922. He had been unable to accept Tessenow's skepticism or his leaning toward the traditional, craftsman's approach to architecture. Wachsmann wanted, as he said, to "incorporate into architecture the revolution that had been ushered in in politics, art, literature, music, and technology. And here, Tessenow had little to offer."

Nevertheless, Tessenow's recommendation allowed Wachsmann to find a position – and to earn his bread in a stricken Berlin – with Leo Nachtlicht, a fashionable architect. "With him I became something like the chief pencil-sharpener", said Wachsmann, "but I was able to develop my drafting skills to perfection. It was there that I first learned how to draw straight and curved lines. My working life concentrated itself on the difference between a 6B and a 6H pencil. But the meaning of precision suddenly dawned on me."

And then, in April of 1923, a decisive meeting occurred. After the premiere of "King Lear" in Berlin's Grosses Schauspielhaus he met Hans Poelzig. Poelzig had designed the sets for the last performance in Max Reinhardt's famous theater on the Schiffbauerdamm. His son, Peter, had been befriended by Wachsmann. As Wachsmann recollects: "Poelzig had an enormous influence on me. He was the kind of man that I would liked to have been. Perhaps there was a kind of inner similarity between us because we immediately felt attracted to each other." With little hesitation, the famous architect accepted Wachsmann as one of his master pupils, among whom were also at that time Egon Eiermann, Richard Paulick, and Julius Posener.

In the commune behind the New Palace in Sans Souci, Poelzig's master pupil found his new home and work place. He also found that there existed a particularly trusting relationship between his teacher and himself. (At least part of the reason for that may have been due to the fact that Wachsmann's grandfather had commissioned and built the factory in Luban, since famous, through which Poelzig had first gained an international reputation.) According to Wachsmann, "Poelzig allowed me considerable freedom and scope. But he regularly discussed my work with me. He was never dogmatic. He was the kind of teacher who encouraged his students and never required that they automatically accept his opinion. But he hated hypocrisy and self-serving compromises."

Through Poelzig, Wachsmann was introduced to the "Scene". He met not only Max Liebermann and Käthe Kollwitz, but also Peter Behrens, from whose offices Gropius and Mies van der Rohe had come. He also met Bruno and Max Taut, Martin Wagner, Ernst May, Wilhelm Kreis, Paul Bonatz and, during a trip they took together to the first Bauhaus Exhibition in Weimar, Jacobus Oud, and Sigfried Giedion, a leading propagandist of modernism, who helped open his eyes to what was happening in the outside world.

The Weimar Exhibition and the thinking of Le Courbusier and Oud reinforced Wachsmann's reluctance to accept traditional architecture, the still-lingering guild approach, as relevant to the 20th century. Wachsmann recalls: "Even while I was still searching, and while Gropius was idealizing craftsmanship, three sentences of Oscar Wilde's impressed me: 'Everything that is machined can be beautiful only when it remains unadorned. Do not try to decorate it. We can only envisage a good machine as elegant – because there is no distinction between strength and beauty.'

4 Poster for the first Bauhaus Exhibition in Weimar. The Exhibition, ordered for financial reasons by the State Government of Thuringia, became an important meeting place for Wachsmann. Here, he met not only the leaders of the Bauhaus, but also Oud and Giedion, with whom he established life-long friendships.

At that time, I didn't know why I took such notice of these thoughts of Wilde. I was also uncertain where the path I had chosen would lead me. It was only clear to me that the workshop of the Bauhaus had no appeal. It was only in 1923, when radio began, that Gropius proclaimed the unity of art and technology as the new way."

But there was little of this to be seen in the Weimar Exhibition. Wachsmann recalled later that, "As late as in 1942", he argued with Gropius whether "it would be in the Bauhaus workshop or in factories and industrial buildings where future architects and builders would learn their skills – or would there even be architects in the future!"

The trip taken together with Oud and Giedion to Weimar proved to be a disturbing factor in the relationship between Poelzig and his master pupil. Wachsmann remained almost another year in the commune, "but then", he related in 1979, "something very strange occurred: I began to be bored with Poelzig. Neither our good relationship nor the freedom he gave me could change that. I wanted to build, and I wanted to build in the modern way. I was interested in standardization, Poelzig was interested in art. His use of materials, which after all culminated in the neo-classical style, had become for me almost poignant. It no longer appealed to me. With Poelzig, I was convinced, art would always dominate. But I was searching for something else, although what that was was still not clear to me."

Abruptly, Wachsmann left Poelzig and the exhilarating atmosphere of Potsdam and its palaces. And Germany. He set out, first to Holland, then France.

But neither with Oud in Rotterdam nor with Le Corbusier in Paris was Wachsmann able to make his fortune. Both would have gladly welcomed him as a colleague, but neither had the financial resources to pay him. Wachsmann's Holland sojourn ended, leaving him impoverished, almost a vagabond, but considerably enriched culturally. Finally, now in Paris, the almost starving bohemian was rescued by his favorite sister, who paid for his return ticket to Berlin.

What Wachsmann brought back with him was the experience of other cultures, meetings with the pioneers of a new architecture, new art, new literature. He formed life-long friendships with some of the leading thinkers of the 20th century.

But once again it was Hans Poelzig who played a decisive role in the life of the difficult young non-conformist. "He summoned me to him and handed over, without a word, a letter and fifty marks. Then he said, in a tone that allowed for no contradiction, 'with this you will go to Niesky. There you will work with Christoph & Unmack. And then we shall see.'"

In this way, Konrad Wachsmann entered the then largest wooden construction company in Europe. There he was introduced to the world of machines, to technologies, to the early beginnings of architectural building, and to the possibilities of mass-production. Out of the carpenter and furniture-maker emerged an engineer, an inventor, and an architect – better, a budding construction scientist – in whose thinking the most up-to-date ideas now took center stage. "Everything that followed, everything that came about in Berlin, New York, Tokyo, Chicago, London, Moscow, Paris, Rome, Zürich, and Warsaw – all that, surprisingly enough, had its beginnings in Niesky, a narrow, provincial village. In this wooden construction company I discovered the way that led me to the turning point in building..."

5 The workplace as laboratory. In the carpentry shop of Christoph and Unmack, AG, in Niesky, Wachsmann discovered the miraculous world of the machine and engineered construction, a discovery which led him to his own 'turning point in building'.

The restless and inventive worker quickly advanced and was soon Christoph & Unmack's chief architect. Soon his innovations could be seen in hundreds, even thousands, of houses, hotels, factories, bridges and train stations.

But it is to chance that Konrad Wachsmann felt he owed perhaps the most significant meeting of his life.

"It was in Niesky, sometime in the spring of 1929. More or less accidentally, I read in the paper a small notice that the city of Berlin wanted to present a country home to Albert Einstein, the Nobel Prize winner, in honor of his 50th birthday. But most important to me was a single sentence: Einstein wanted a wooden house. When I read that I knew – Konrad Wachsmann would build that house!"

Without hesitation, he traveled to Berlin to present himself as a specialist in the building of wooden houses to the most famous man of the century. But it was not the physicist, but his wife, Elsa, who greeted Wachsmann at the door of Haberlandstrasse 5. And it was not the architect, but rather the impressive Christoph & Unmack company car in which he arrived that convinced her that he was the only right man for the job – despite the fact that Gropius, Mies van der Rohe, the Tauts, and Martin Wagner all lived and worked in Berlin. Le Corbusier would probably even have traveled from Paris to try for the commission of the century.

6 Albert Einstein and Konrad Wachsmann on the terrace of Einstein's country home in Caputh. "Everyone was impressed and happy." The photo was taken by Margot Einstein in 1929.

It was also Elsa Einstein who interpreted for the young architect her husband's idea of his dream house. During his last visit to Caputh in 1979, Wachsmann reminisced: "Until today I am not certain whether they were her, or her husband's, wishes. A brown-stained wooden house with small, white, French windows, a overhanging roof of red tiles. Except for a large living room with a fireplace, all the other rooms were to be small. Most importantly, Einstein's bedroom and his work room should be set well apart from the rest of the house. He snored loudly. And he needed quiet for his work. In addition, several terraces were required so that they could spend as much time as possible outdoors."

The country home as we know it is without a doubt a compromise between the wishes of its owner and the ideas of its architect. Einstein got his French windows, his red-tiled roof, and his isolated work and bed rooms. Wachsmann got his living area, spacious and open, roofed with enormous terraces.

Almost before it was finished, the Einstein house became a public attraction. Following the journalists came the curious, who simply wanted to see how the famous man and his family lived.

As Margot Einstein, in 1985, remembered it: "When the house was completed, everyone was impressed and happy. Konrad Wachsmann had built for us a really attractive, well put together house. There was nothing luxurious about it, it stood out rather through its under-statement which allowed its clean lines and the beautiful woods – whose odor lin-

7 Paradise on Lake Havel. Albert Einstein in his house in Caputh. Photographed by Konrad Wachsmann.

gered for a long while – to fit so wonderfully into the surroundings. The whole family was enchanted by it."

Above all, Albert Einstein. In one of the few extant letters from Albert Einstein to Konrad Wachsmann, Einstein wrote (on July 25, 1931, from Caputh): "I can only tell you that nowhere else and in no other house have I felt so well and so much at home. And all our visitors, who understand more about architecture than I do, have always expressed their enthusiastic admiration for what has been achieved."

And there was no shortage of visitors to Caputh. Although a guest book was only sporadically in use, one knows today who were among the invited – and uninvited – guests. Max von Laue, Tagore, Chaim Weizmann, Alfred Kerr, Erich Kleiber, Hermann Struck – they were among the few of the prominent visitors who were able more or less intelligently to inscribe themselves in the guest book. Einstein, it should be known, asked that inscriptions be "Verses pure and sweet, like those of sublime poets". Apparently neither Max Liebermann nor Hans Poelzig wanted to respond to the challenge; they came only so that they could judge Wachsmann's work.

There are also no entries from such visitors as Anna Seghers, Heinrich Mann, Käthe Kollwitz, the Russian director Natalia Saz and her countrymen Lunacharski and Leonid Pasternak, father of the famous Boris Pasternak, the author of Doctor Zhivago. Neither Stefan Zweig nor Gerhard Hauptmann seemed inclined to versify. No entries either from the famous attorney, Rudolf Olden, or the publisher, Samuel Fischer. Nothing either from Einstein's colleagues in the scientific world, all of whom either were or would become Nobel Prize winners: Max Planck, Walter Nernst, Fritz Haber, Max Born, Erwin Schrödinger, among others.

One frequent and welcome visitor to Caputh was Konrad Wachsmann himself.

In 1929, he opened his own office in Berlin. Encouraged by the facilities he needed to further develop his ideas about industrializing building construction processes, he had left Christoph & Unmack. Although in Niesky his working environment had been close to ideal, and although he had been free to experiment more or less as he pleased, he felt the technical facilities available there placed ever increasing limitations on his work. His venture, however, coming just at the time of the Great Depression, came to a quick end.

Perhaps more than he would have wished, therefore, Wachsmann now had time for long walks with Einstein, to go sailing with him, and to join in lively discussions with other visitors to Caputh. During these critical times in the life of the Weimar Republic, the Nobel Prize laureate's country home had become a meeting place, where conversations ranged over the gamut of the important political, cultural, and scientific questions of the day.

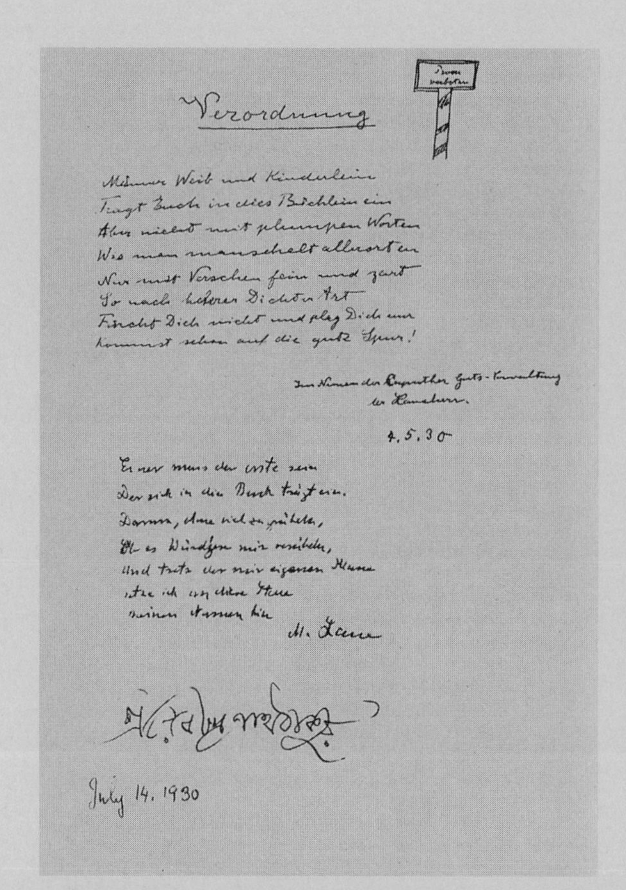

Verordnung

Männer Weib und Kinderlein
Tragt Euch in dies Büchlein ein
Aber nicht mit plumpen Worten
Wie man manscht allerorten
Nur mit Verschen fein und zart
So nach höherer Dichter Art
Fürcht Dich nicht und plag Dich nur
kommst schon auf die gute Spur!

Im Namen der Caputher Guts-Verwaltung
der Hausherr.

4.5.30

M. Laue

July 14, 1930

8 Entries in the Einstein house guest book. Einstein as host required "verses pure and sweet".

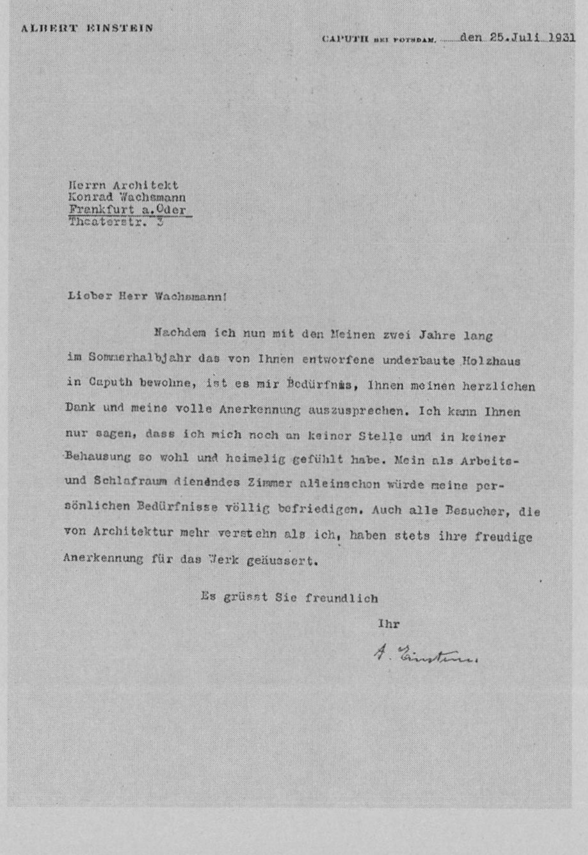

ALBERT EINSTEIN

CAPUTH BEI POTSDAM, den 25.Juli 1931

Herrn Architekt
Konrad Wachsmann
Frankfurt a.Oder
Theaterstr. 3

Lieber Herr Wachsmann!

Nachdem ich nun mit den Meinen zwei Jahre lang im Sommerhalbjahr das von Ihnen entworfene underbaute Holzhaus in Caputh bewohne, ist es mir Bedürfnis, Ihnen meinen herzlichen Dank und meine volle Anerkennung auszusprechen. Ich kann Ihnen nur sagen, dass ich mich noch an keiner Stelle und in keiner Behausung so wohl und heimelig gefühlt habe. Mein als Arbeits- und Schlafraum dienendes Zimmer alleinschon würde meine persönlichen Bedürfnisse völlig befriedigen. Auch alle Besucher, die von Architektur mehr verstehn als ich, haben stets ihre freudige Anerkennung für das Werk geäussert.

Es grüsst Sie freundlich

Ihr

A. Einstein.

9 "...enthusiastic admiration for what has been achieved". In a letter from Einstein to Wachsmann.

Although Wachsmann had established himself and been recognized as an expert in the field of wooden construction before the Einstein house, after 1929 and by virtue of his association with Einstein he entered into the international spotlight. For the Berlin publishing house of Ernst Wasmuth, this sudden prominence was reason enough to secure Wachsmann as one of their authors.

Wachsmann's *Holzhausbau* was published in October, 1930. (And is now, fortunately, being re-issued.)

In his book Wachsmann aimed to describe the "transformation in the way we view building" for a broad, general public. In the life of the architect, now also turned author, the book would shortly take on a dramatic if unexpected role.

Clearly, the publisher Wasmuth, and his publicity department had calculated not only on the sudden fame of Wachsmann himself, but also on the appeal of the name Einstein, now a household word. Einstein, who shied away from all publicity, saw with some regret that his sanctuary in Caputh – already an all-too-open secret – was also now to be described in a book. Despite his craving for privacy, he gave Wachsmann permission to include material on his refuge, an act of generosity attributable to the great attachment that had developed between him, the aging Nobel Prize winner, and his young protégé.

But even the success of the book could not alter the increasingly difficult situation in which Wachsmann now found himself.

Not only the economic situation but also the political climate led Wachsmann, in the summer of 1932, to compete for the prestigious Rome Prize of the Prussian Academy of Arts. Wachsmann hoped thereby not only to distance himself from the unrest in Germany and a climate of opinion that had become for him increasingly distasteful, but also to find fresh intellectual and creative inspiration. Furthermore, he had tired of the hectic life in Berlin and the endless discussions in the Romanische Café and at Einstein's home.

He had come to a critical turning point in his life.

His book, *Holzhausbau* in hand, he presented himself to Heinrich Haslinde, the State Secretary in the Prussian Ministry of Science, Art, and Education. It was the book that convinced the State Secretary, his officials, and also Max Liebermann, the President of the Prussian Academy of Arts (who was already well acquainted with Wachsmann) that Wachsmann was "a talent with outstanding architectural sensitivity". The doors of the German Academy in Rome, at the Villa Massimo, were opened.

When the proud winner of the Rome Prize went to say his farewells in Caputh, Einstein was in the process of preparing himself for his annual working visit to the United States. Neither of them could have guessed that this was to be their last meeting in Caputh. After Hitler's take-over and the Holocaust, Einstein was never again to set foot on German soil.

In the late autumn of 1932, Konrad Wachsmann set out for Rome, traveling via Vicenza, Padua, Ferrara, Bologna, Florence, Sienna, and Assisi. He arrived in Rome in December.

He was overwhelmed by his surroundings in the Villa Massimo. He wrote his mother: "When I arrived at 9:30 in the morning in the park of the Villa Massimo I was almost paralyzed. One felt, more or less suddenly, as though one had been transformed into a prince, and the first thought was: how will it be when one has to leave this place. Here, one can become spoiled for one's entire life. Only a millionaire could afford this."

Wachsmann was well satisfied with his atelier in the marvelous park of the Villa Massimo,

10 A welcome guest. Konrad Wachsmann in Caputh, summer 1932.

and he quickly established a friendship with the director and his wife, Herbert and Erica Gericke. But his good fortune was to be short-lived. Despite his strenuous efforts, Gericke was not able to insulate the Academy from events in Germany. The arts were not immune to National Socialism.

First, there was the infamous book burning of May 10, 1933, and then a fight in which the Jewish painter, Felix Nussbaum, was badly hurt by Hans Hubertus, Count of Merveldt, a Nazi of few talents. Events came to a head when Wachsmann, after violent disagreements, confronted Arno Breker in a pistol duel, witnessed by Erica Gericke and reported by Herbert Tucholski in the *Weltbühne*.

All enough reason for Wachsmann to relinquish his Rome Prize. Wachsmann: "After Goebbels' insane attack on the culture and the spirit of our people, I had no other choice. It was impossible for me to feel myself part of a state that had so openly reverted to barbarism."

There followed a hair-raising flight from fascist Italy, internment and the confusion of war in France. And then, through the efforts of his friends Einstein and Gropius, and countless others who had already escaped to the United States, Wachsmann was also able to find refuge there. At almost the same time, his mother, his sister and her son – who had all remained in Germany – were brutally killed in a concentration camp.

That Wachsmann together with Ida Chagall and many others were able at almost the last minute to escape the SS and Gestapo executioners was – it should be pointed out – in no way due to Wachsmann's professional prominence. The US also practiced a form of selectivity. With some exceptions, neither doctors, lawyers, accountants, or architects counted among those chosen for rescue. It was authors, artists, and scientists, who, by special instructions from President Roosevelt, were viewed favorably by American immigration authorities.

And now *Holzhausbau* played a critical role. Because Einstein and Gropius remembered it – and at just the right moment – Wachsmann escaped the concentration camp, and presumably the gas chambers. It was not Wachsmann the architect who was able to survive, but Wachsmann the author.

Soon after his arrival in New York, it was Wachsmann the architect, however, who was able to realize his version of the American dream. Together with Gropius, the impoverished refugee founded the General Panel Corporation, the first factory to mass-produce pre-

11 Waiting in exile. Konrad Wachsmann in Italian exile, Venice, 1935. The German ambassador, von Hassell (later to become a member of the group involved in the abortive attempt to assassinate Hitler) protected Wachsmann against the Gestapo and the Italian Secret Police.

fabricated building elements. Five unskilled workers could now in eight hours construct a house that was not only affordable, but could be immediately occupied.

The General Panel House, which entered into American history as the "peace-time dream house", was grounded in the principles that Wachsmann had already developed in Germany and during his years of exile in Italy and France.

There are many myths concerning the sudden collapse of the General Panel Corporation, in which many parties in the US and also many German refugees had participated – including Sigfried Giedion, its foremost champion. It remains certain, however, that with the General

12 Hope against hope. Wachsmann's sister, Charlotte (right), and her son shortly before their deportation to Riga. The family hoped that 'Uncle Milch', Göring's deputy, would be able to protect them. Erhard Milch, a Jew, had financed Wachsmann's studies and had gained his reputation as a deputy director of Lufthansa. The Nürnberg Tribunal found him guilty of war crimes.

14 Friends, partners, colleagues. Mies van der Rohe and Konrad Wachsmann in Chicago. The New Bauhaus, the Institute of Design, and the Illinois Institute of Technology set new international standards in architectural research, teaching, and practice.

Panel System a milestone was achieved in the field of pre-fabricated construction.

Shortly thereafter, new horizons were opened for Konrad Wachsmann – in research and teaching. With his appointment as professor at the Chicago Institute of Design, the successor institution to Moholy-Nagy's controversial "New Bauhaus", Wachsmann, together with Mies van der Rohe, developed countless projects. They broke fresh ground; once again Wachsmann established himself as a pioneer in the field of architectural engineering.

13 Walter Gropius and Konrad Wachsmann at a construction site of the General Panel Corporation. "A finished house in only 8 hours."

Among his projects was the construction of hangars out of pre-fabricated tubular steel elements for the US Air Force, that in size far exceeded anything yet constructed. The system has since been utilized in almost every airport in the world, and particularly in buildings with extensive over-hanging roofing.

This system was developed over a three-year period of experimentation, used by Wachsmann to elaborate a revolutionary method of team work between teacher and student. New standards were established which altered the teaching methods in schools of architecture everywhere as well as the planning processes of most architectural firms around the world. Wachsmann described this in his book Turning Point of Building, even now accepted as a standard work. His principles have since been taken up

15 Two Chicagoans at Max Bill's Institute for Design in Ulm, Otto Kolb and Konrad Wachsmann. Kolb, Alfred Roth's office manager for many years, came to Chicago in 1948 through the efforts of Giedion. "Switzerland is sending you the best it has to offer". Giedion told his friends, Gropius, Mies van der Rohe, and Chermayeff.

and revised to meet contemporary standards and requirements by such prize-winning and widely-respected architects as Foster, Rogers, Piano, Otto, and Hochstrasser.

By the time of his death, in Los Angeles on November 25, 1980, Konrad Wachsmann, too, had received many distinguished awards.

One year before he died, Wachsmann revisited the Einstein home in Caputh. With his help, it had been restored. A costly project, accomplished through the efforts of the astrophysicist Jürgen Treder. Due solely to his efforts, the house was saved from falling into total ruin, and has been preserved to be used for fitting purposes.

It was only after the fall of the Berlin Wall that Judith Wachsmann, now a designer and professor of art in Los Angeles, came to Caputh to see the Einstein House, the house on whose foundations the life-long friendship between Albert Einstein and his architect had been grounded.

16 "...ideas and projects, always far in advance of their time, often utopian". A model of the City Hall in California City, one of Wachsmann's last projects.

17 He "fascinated an entire generation of architects, all over the world". Konrad Wachsmann in Austria.

In the house, now being maintained as a museum, she discovered a copy of *Holzhausbau*. When it was first published, Konrad Wachsmann had dedicated the book to "his loyal colleague, Walter Klausch". Klausch had worked as an engineer on the construction of the Einstein house, and, as a member of the Communist Party, had become in 1933 one of the first victims of the Nazi terror.

And now, in sad conclusion: Since the reunification of the two so disparate German republics, the Einstein House in Caputh stands under 24-hour guard – to prevent the new German Nazis from setting it to flames, lighting yet another nihilistic ideology.

18 Konrad Wachsmann, Michael Grüning, and Georg Opitz at the Bauhaus in Dessau, 1979. "Pius was our god and Pia his deputy, and the Bauhaus embraced us all. But for me 'he', I believe, did more than anyone else. And now I stand here, with all my guilt feelings, and can only admire the Bauhaus."

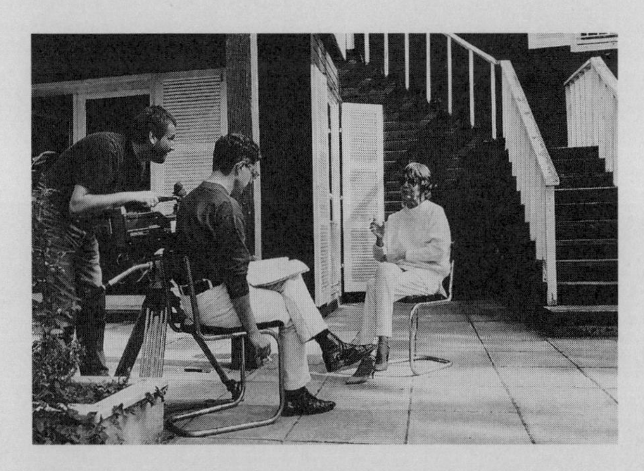

19 "Led by Einstein to the outer limits of our knowledge." Judith Wachsmann in Caputh, fall 1992, during the filming of the book, "Ein Haus für Albert Einstein". The film was a production of the German television channel, ZDF, and the European cultural channel, 'Arte'. Photo by Roger Melis.

Photo credits: 1, 2, 4, 6, 7, 8, 10, 11, 13, 16, 17, 18 from Michael Grüning, *Der Wachsmann-Report – Auskünfte eines Architekten* and *Ein Haus für Albert Einstein*, Published by Carl Hanser, Munich. Photos 3, 5, 9, 12, 14, 15 are from copies in the Konrad Wachsmann Archive, Huntington Library, San Marino, California. The originals are in the archives of Verlag der Nation, Berlin, collected by Michael Grüning. Photo 19, Roger Melis.
For photo assistance, the authors and publisher wish to thank Judith Wachsmann (Glendale, Arizona) and Ray Wachsmann (Los Angeles, California), as well as Roger Melis (Berlin).

Christian Sumi

Building the Wooden House Today

It was during a trip to Copenhagen in 1984, to see the work of Arne Jakobsen, that Marianne Burkhalter and I, rummaging through old book shops, came across Konrad Wachsmann's *Holzhausbau (Building the Wooden House)*. We leafed through it many times thereafter (usually back to front) without, admittedly, really reading or studying it closely. We were, instead, fascinated by several of the photos and picture series. For us they provided an illustrative lesson in design and form and in perception of form, but they also showed an appealing informal and non-dogmatic approach to architecture. These were concepts that had, since the '80s, attracted us in our search for a more immediate form of architecture, one beyond rigid and inflexible terminology and classification.

The Subtlety of the Illustrations

In the Berlin Public Transport Authority Building, built in 1928 (page 121, top), the modern, thin outer casing is set against a traditional, symmetrical construction, and a wide-eaved, traditional roof, to contrast the distinctive characteristics of the traditional with the modern. This is further emphasized by small windows, set slightly off-center – a subtle device to indicate that this is a panel construction, using "panel joints without stripping". Emphasis is further added by the composition of the photo itself, which, in almost contrapuntal form, shows a figure with cigarette, possibly Wachsmann himself, walking away from the building. The photo on the opposite page adds an almost surrealistic effect, showing the same person on the porch and a figure in the window placed between railway tracks and overhead electrical wires. These two photos, taken together with the view of the open, deserted offices, produce an effect very close to that of a scene from a suspenseful photo thriller.

1 Konrad Wachsmann, Office building of the Berlin Public Transport Authority.

The opposed photos (pages 88 and 89) of the Earth Sciences Institute in Ratibor (1928) emphasize the relationship between silhouette and building – first photo the two-dimensional, the second showing depth. Almost ironic is Wachsmann's caption referring to "the extensive terracing on the building ... provided for the placement of scientific instruments" – which does not appear in the photos at all. The terraces have been deftly integrated into the building and give to the pro-

2,3 Konrad Wachsmann, Earth Sciences Institute in Ratibor, Main view and north side.

ject its pleasing compactness. This is in contrast to the then-current emphasis on presenting in detail individual building elements. A subtle indication of the terraces is given in the photo left through the exposed rain spouting, which is solidly in the surrealistic tradition of the "objets absents". On the facing photo, the spouting is no longer visible. One's attention is also drawn to the special montage effect of the photos. By separating an overall view in two, with a depth photo on the left and a flat, two-dimensional photo on the right, plus a scarcely noticeable change in perspective, and the placement of the photos in the middle of the book, the photos can be folded together so that the vanishing point of the two photos is spread apart as widely as possible. Joining and separating the photos produces an artificial unity of great, almost irritating, visual strength.

In the following photos (pages 90-91), the spouting becomes a major element of the façade. The spouting, together with the corners and the roof edging, divide the building elements, their façades further divided by the windows with their color high-lighting. Wachsmann provides no blue prints; it is almost impossible to conceptualize the project in its entirety.

4,5 Konrad Wachsmann, Earth Sciences Institute in Ratibor, North side and rear view.

The illustration section of the book begins on page 45 with the Children's Convalescent Home in Spremberg, a frame construction and the first large project completed by Wachsmann, in 1927, for the Christoph and Unmack Company. The illustrations are laid out on facing pages, using vertical plates on pages 46 – 47, 54 – 55, horizontal plates on pages 48 – 49, 50 – 51 and so on, with accompanying smaller photos on the lower part of the page. Shown in this form, the project illustrates the difficult relationship between the layout design and its realization. The underlying cross design of the layout is never in its totality – as a cross – discernible. It is only by the use of the small photos of models, presented in footnote form, that the external and internal corners can be understood as forming, together with the raised playground terrace, a cross (pages 48-51). In particular, the photos on pages 46 and 47, showing windows mounted on the outside frame (page 46) and the inside frame (page 47) are used to illustrate the relationship between façade and windows, between rhythm and structure.

6 Konrad Wachsmann, Children's Convalescent House in Spremberg.

Contrasting heights emphasize the interrelationships between the worlds of the adult and the child in the photo on page 59 of the washroom in the Convalescent Home. The high-standing radiator with the faucets arranged above – which, as Wachsmann points out in the caption, are to be regulated by the staff – are contrasted with the low benches, sinks, and simple faucets for the children's use. This is further contrasted in the reflections that appear in the spilled water on the floor, a device used to show also the quality of the floor itself. The towel racks, placed at differing heights on the far wall, are used to provide a kind of scale against which the various heights can be measured. The doctrine of functionalism – the easily maintained floor, and a well-lighted, well-ventilated room with generous working space, shown on the opposite page, is always tempered by small, everyday details. A subtle counterpoint to the new formalism and photographic social realism.

So it was that the suggestiveness of the illustrations in the second part of this book struck us in much the same way as Giedion's use of captions "for the fast reader" in his *Space, Time and Architecture*. It is clear that by "fast readers" Giedion meant architects. For us, as "fast readers", not only the narrative but also the contrast of differing architectural conventions proved fascinating. A further example in the illustrations is the placement of van de Velde's thatched, "traditional" country home (page 114) between photos of Bruno Paul's (page 115) and Scharoun's (pages 112, 113) modern designs, (all constructed using the panel method.) Or, on pages 98 and 99 (and fully described on page 23), Hans Poelzig's planked and wood-encased home opposed to the plaster frame construction of Paul Schmitthenner.

Surprising is the absence of actual blueprints, so unlike other contemporary publications, for instance Stolper's *Bauen in Holz* (Building with Wood), (Hoffmann, Stuttgart, 1933), or the monographs appearing in *Moderne Bauformen*, (Modern Architecture), November 1933, which document projects in great detail, using building plans.

Wachsmann preferred to list the basic principles of wooden house construction, and to illustrate them in as comprehensive and suggestive a way as possible. His central theme is not the exhaustive listing of all possibilities in the manner of a building manual, but rather to outline a process, and to sketch broadly a variety of methods and their characteristics.

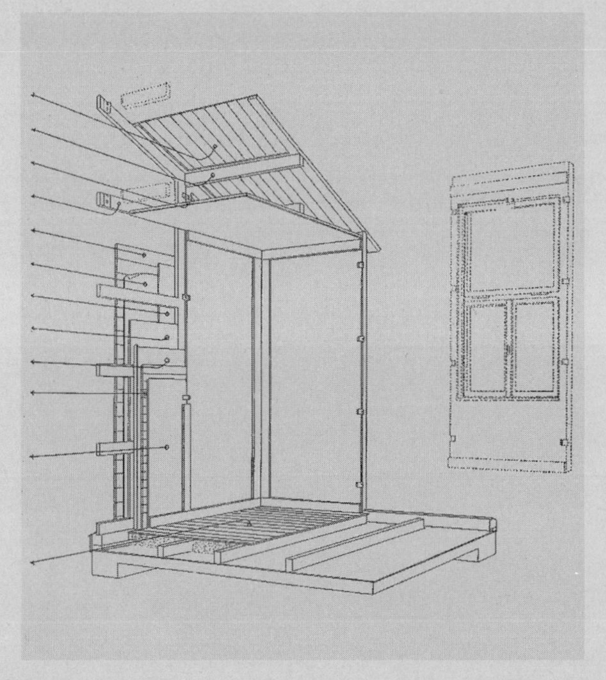

7 Konrad Wachsmann, Diagram of Christoph & Unmack AG's panel method, Niesky, ca. 1930.

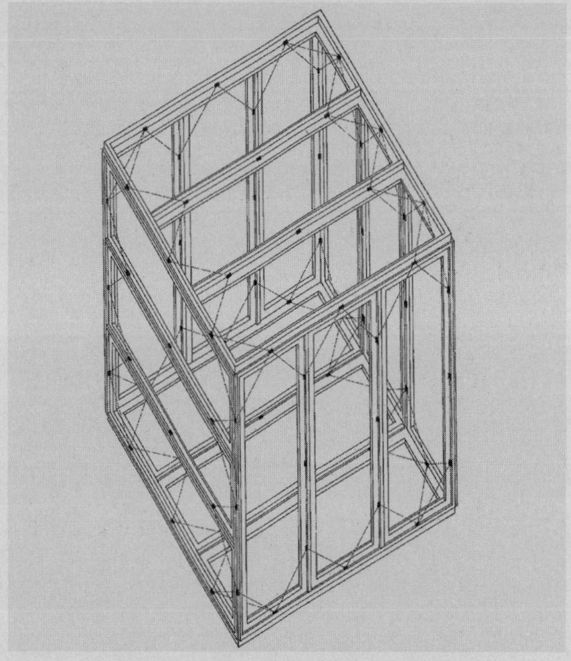

8 Konrad Wachsmann and Walter Gropius, Diagram of the General Panel System, USA, from 1941.

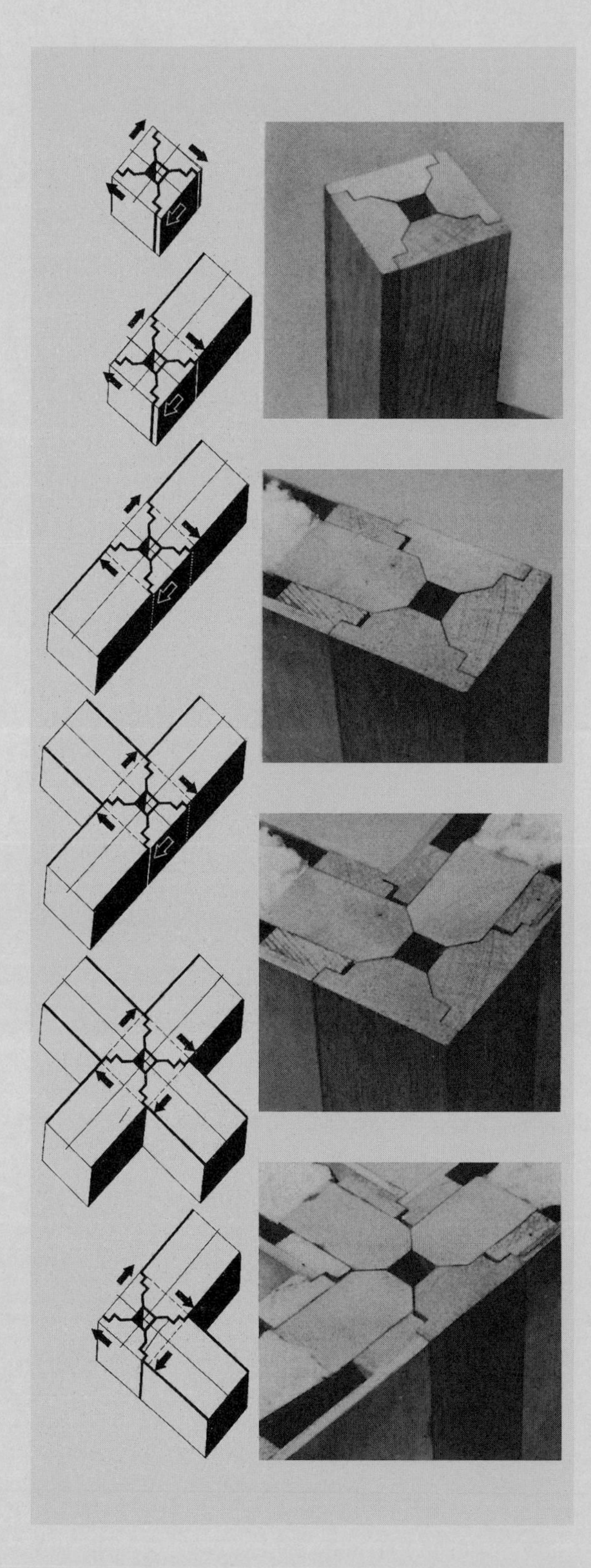

9 Konrad Wachsmann and Walter Gropius, Corner joining in the General Panel System, USA, from 1941.

From the Panel Method (1930) to the "General Panel System" (1941)

"Today, the wooden house is produced by machines in factories, not by the craftsman in his shop." With this programmatic statement Wachsmann introduces the first part of *Building the Wooden House* and, at the same time, brings the panel method to the forefront of his discussion. It is this method which dominated all of his later projects, including the "General Panel System" that, beginning in the United States in 1941, he developed together with Walter Gropius. The system was described by Wachsmann in his book entitled *Wendepunkt im Bauen* (Turning Point of Building), published in 1959. The book has since then become the manual for all aspects of building with wood and pre-fabrication. It has served as a guide for the work we have been engaged in during the past several years.

In constructing the walls, the panel method is basically the same as the wood frame method. As in the American platform system, the dimensions of the supporting cross-sections match the boards used, but are laid horizontally rather than vertically to frame the wall sections. Top and bottom plates provide stability so that each panel can be handled singly. These are then joined together to form a wall frame, using clamps at the corners to provide stability.

This kind of panel formed the basis for the General Panel System, which was to be developed further, however, in three areas:

First, the frame dimensions are calculated on the basis of the actual amount of support required, and at the joints are sized to fit rod-like posts.

Second, Wachsmann's main line of experimentation was in the direction of developing the three-dimensional iron joint. It was here that Wachsmann showed himself at his most creative. The wedge-shaped wooden elements are arranged asymmetrically to allow for the fitting together of the wooden framing and form, therefore, an angled, three-dimensional joint. The frames and panels are then fastened together with iron joints set on the inside of the frames.

Third, while the floor and ceiling of the panel system is constructed in the conventional way, they are integrated in the General Panel System. Wachsmann developed a frame which could be used as either a wall or floor element even though the support required of these elements varies in terms of load-bearing, stiffness, and lateral stability. The floor elements are laid on a frame-like sub-floor and are non load-bearing.

10 Frank Lloyd Wright, Exterior view of Jacob's House, 1936.

With this, the transition is completed between the method using pre-fabricated individual elements and the modular system in which the joining principle (rotation) becomes the dominating architectural principle while the actual method involved (the use of the iron

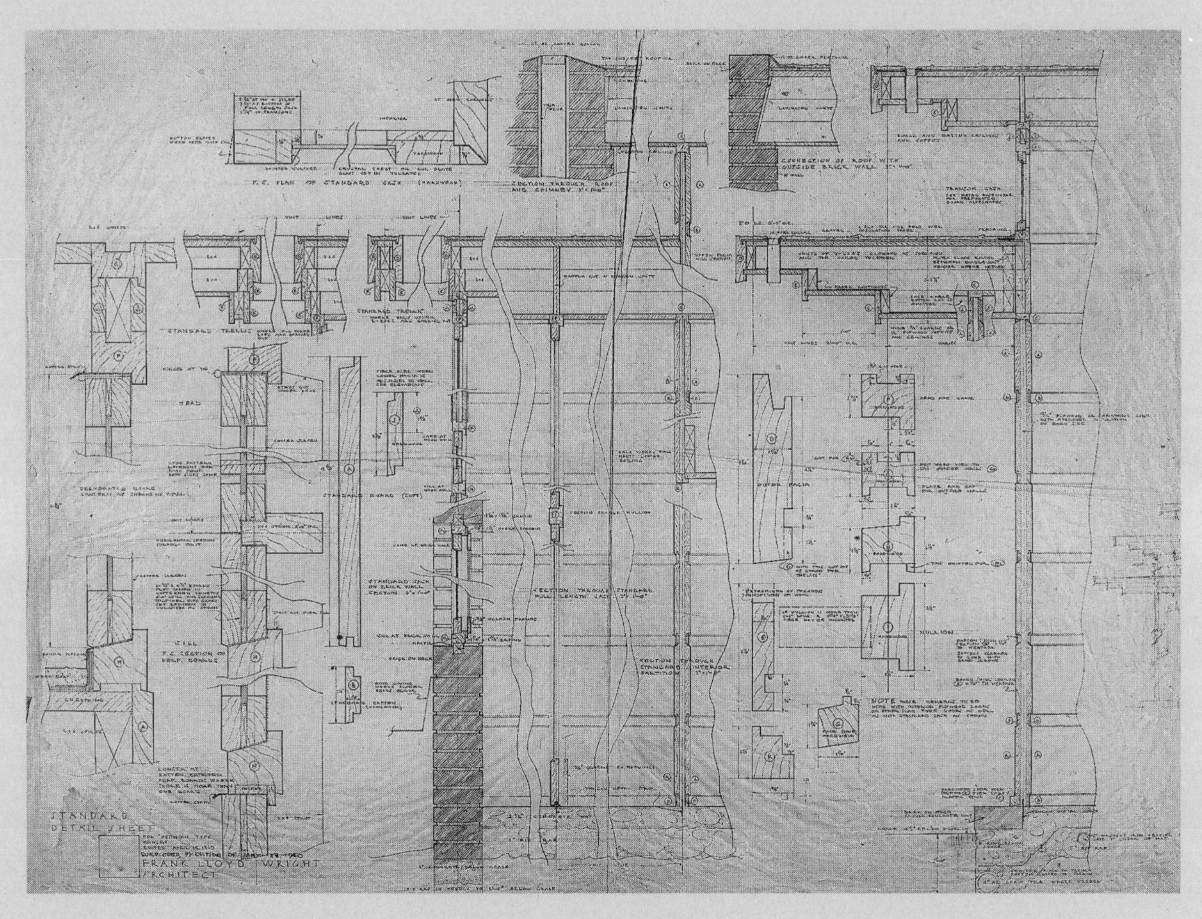

11 Frank Lloyd Wright, Usonian Houses, Standard Detail Sheet, 23 May 1940.

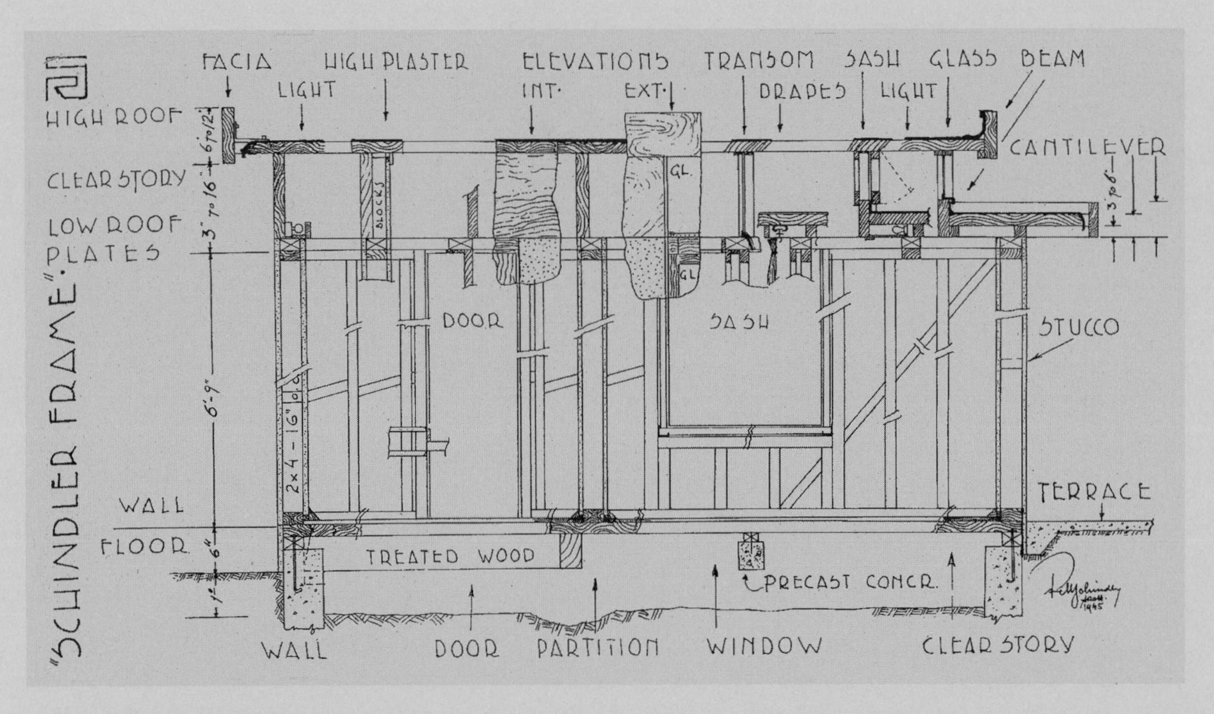

12 Schindler Frame, 1945.

joint) disappears from sight. At the same time, traditional concepts such as load and non-load-bearing, upper and lower, horizontality and verticality become ever more subordinate to an abstract, mathematical view of spatial relationships.

Frank Lloyd Wright's Usonian Houses and the Schindler Frame

Immediately before and after the development of the General Panel System other architects in the United States, such as Frank Lloyd Wright and Rudolf M. Schindler were also moving in the direction of the modern, prefabricated wooden house. Wright's Usonian houses were based on the concept of a building block consisting of different components. A concrete slab is the basic unit into which 4´ x 4´ frames are inserted topped by a 4´ x 8´ plywood section, used as ceiling. Floor heating, a novelty at the time, is built into the floor slabs. The walls are formed of core vertical plywood sections faced on both sides with paneling and molding outlining 13 inch frames. Insulation paper is placed between the plywood and the framing. The roof construction is, together with the various skylights, a part of this unit and consists of three vertical 2˝ x 4˝ boards with upper and lower molding. To span larger distances, particularly near the sky lighting, steel reinforcing is placed out of sight within the wooden elements. Stiffening is provided through the large brick or natural stone core bathroom, kitchen and fireplace units.

A distinguishing feature of the Schindler Frame lies in the division of the wall into a lower, door-level unit, consisting of a simple, planked 2˝ x 4˝ assembly, and a space-saving roof construction, consisting of a level above the door lintel, a board panel extending through a clear storey, and the roof. This open, space-efficient level serves various functions, including ventilation, various skylighting, direct and indirect lighting, the placement of cur-

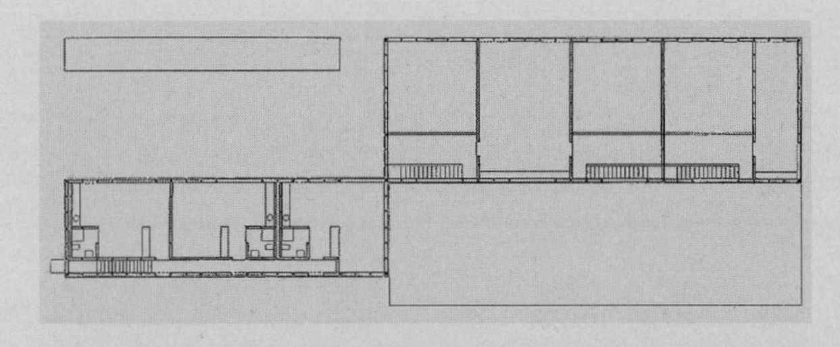

13,14 Burkhalter and Sumi, Kindergarten in Lustenau, panel method, Vorarlberg, 1993. The cross formation of the layout is not discernible, leading to a playful kind of puzzle arrangement between the parts and the whole. The serial arrangement of vertical and horizontal windows emphasizes the contrast.

15 Modern application of the tongue and groove technique. The veneered posts and beams are so constructed as to form the inner skeleton. Tongued rods placed between the posts and windows absorb the shifting between casing and load-bearing elements.

tain rods, etc. The floor is laid on parallel joists supported, in the case of larger areas, with concrete footing.

Homogeneity versus Individuality

The building block versus the building system; the focus on individual building units versus integrated, total pre-fabrication of the whole building – these approaches characterize the difference between Wright and Wachsmann. Particularly in the way the wood is handled, a kind of hybrid, low-tech style is contrasted with a monolithic, high-tech style. The contrast is also apparent in the way space is utilized: architecturally determined, fixed spaces are opposed to a more neutral, indeterminate fixing of space. It is within the extremes established by Wachsmann's General Panel System and Wright's Usonian houses that pre-fabrication is fixed.

Konrad Wachsmann was able to give the idea of abstraction a place in the architectural debate, thereby advanced pre-fabrication, and in particular its process, to the point of becoming the driving force in the post-war, international style.

Nonetheless, we think that pre-fabrication, with its homogeneous implications, is in contradiction to our contemporary notions of production – "lean" and "just in time". It is too ponderous and can and will not react to varying requirements or changes in the market (- these are perhaps also reasons why the General Panel System foundered in the 1940s). Contemporary wooden building looks now not to a reductionism seeking a universal panel, but to the development of an appropriate multitude of different elements, responsive to a variety of requirements, including ecological ones.

The ways, however, of handling wood – beveling, notching, the problem of joining, all of which

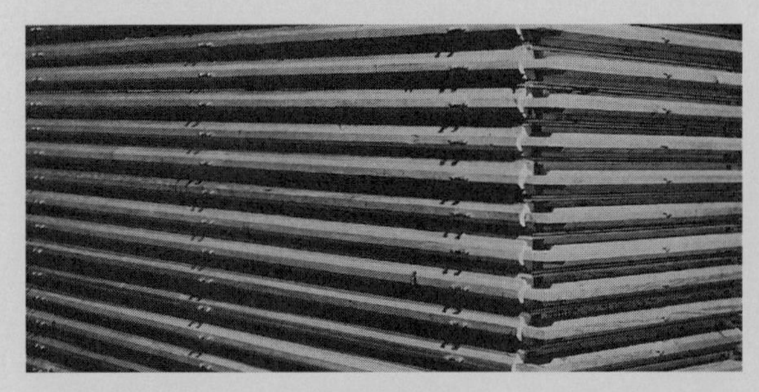

16 Stacked General Panel System frames.

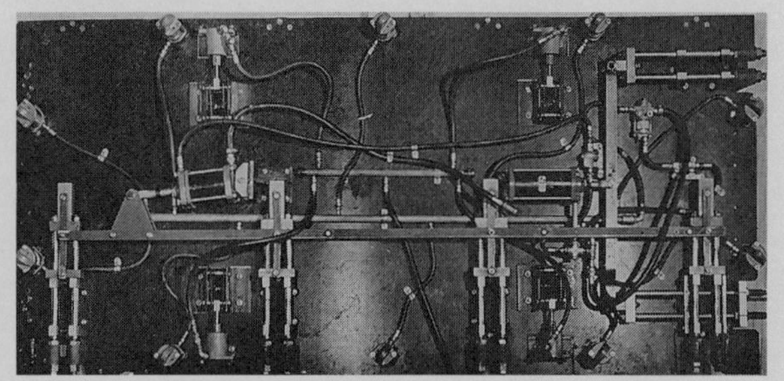

17 A rear view of the machine used in frame production.

Wachsmann dealt with almost magically – are today of as much concern as before. Today, with even the simplest milling machines qualitatively excellent work with wood is possible, as Wright's Usonian houses demonstrate so well. Pre-fabrication has thereby been freed from the somewhat overwhelming influence of high-tech; the way to a more complex, more varied application of building technology has been opened.

The mixing together of style and material – the trend towards a universal style – will, as in the case of the modular wood system derived from concrete building systems, undoubtedly have an impact on wooden building, depriving it of some of its uniqueness. We are unperturberd by this development and even welcome it. We are of the opinion that this will introduce into pre-fabrication architectural and spatial considerations somewhat akin to what Wright accomplished with his individualization and differentiation of building elements. The heavy bottom plates, the horizontally layered walls, the heavily cantilevered roof – all are means by which he emphasized the sinuous strength and the elegance of spatial relationships in his buildings.

Concept as Perception

In Wachsmann's later work in the United States, system became his overriding concern. Three-dimensional frame constructions and experiments with spatial arrangements were at the center of his architectural work, and especially that of his students. At the expense of the personal and individualistic, Wachsmann sought the universal, the universally applicable; his overriding goal became the "pure " design. Alternatives were ignored, unlike with, for example, Le Corbusier, who used precisely that conflict between function and style as his inspiration. With Wachsmann, concept became equivalent to perception: the thought is the building is what is seen – a point of view which brought Wachsmann close to conceptual or minimalist art. His was an architecture that gave over to the inner structure, the engineering, architectural and aesthetic value. The result is a kind of deconstruction of architecture, which paradoxically, in its best examples, manifests almost anonymous strengths.

The world of pure, rod-like designs stands in opposition to the world of pneumatic mechanisms, visible in the machines which this style seeks to replicate, metaphors for softness and synergy. The French architect and builder, Jean Prouvée, would use this way of working with his materials – bending, folding, reshaping – as the starting point for his work.

It was especially through the work of English architects in the '80s that Wachsmann's ideas were revived and given new dimensions. The massive hangars built by Wachsmann in the '50s, that stand in contrast to Buckminster Fuller's attempts to bring heaven and earth together with his geodesic domes, foreshadowed this development. Precisely the enormous new projects of Foster and Piano, in collaboration with engineering firms such as Ove Arup and Partners, demonstrate how the apparently diverse positions taken by Wachsmann, Prouvée, and Fuller, can be brought into a useful symbiosis, lying somewhere between high and low tech.

With his endless curiosity concerning the technical problems of building, and with his penetrating questioning, Konrad Wachsmann set new standards for contemporary industrial construction. His *Building the Wooden House* appeared at the beginning of his search. It is exactly because of this tentative and exploratory spirit that we are drawn to it over and over

Photo credits: 1 – 7 Konrad Wachsmann, *Holzhausbau*, Berlin 1930, p. 121, p.88/89, p. 90/91, p. 47, p. 109; 8,9 Konrad Wachsmann, *Wendepunkt im Bauen*, Wiesbaden 1959, p. 144, p. 141, p. 142; 10 Daniel Treiber, *Frank Lloyd Wright*, Basel, Boston, Berlin 1988, p. 87; 11 Frank Lloyd Wright Foundation, Nr. 3813.009; p. 172 12 RM Schindler, edited by Lionel March and Judith Sheine, New York 1993, p. 63; 13 Burkhalter and Sumi, Zürich; 14 Heinrich Helfenstein, Zürich; 15 Burkhalter and Sumi, Zürich; 16,17 Wachsmann, *Wendepunkt im Bauen*, p. 146, p. 150

KONRAD WACHSMANN

BUILDING THE WOODEN HOUSE

TECHNIQUE AND DESIGN

ERNST WASMUTH VERLAG AG., BERLIN

DEDICATED TO FRIEDRICH ABEL

TABLE OF CONTENTS

Many of the illustrations used in this book have been generously made
available by the following firms:
Christoph & Unmack A.G., Niesky, Saxony
Deutsche Werkstätten A.G., Hellerau
Carl Tuchscherer A.G., Berlin
For many years these firms have carried on pioneer work
in the wood-building industry.
(This volume was originally printed by Otto Elsner K.G., Berlin. Original binding
by J. Fikentscher, Leipzig.)

TODAY, THE WOODEN HOUSE is produced by machines in factories, not by the craftsman in his shop. A traditional, highly-developed craft has evolved into a modern machine technology; new applications and new forms are being developed. Wood simply as lumber – as traditionally used by the carpenter – is no longer able to meet today's requirements. But as a standardized, machine-produced, pre-fabricated product wood can compete in terms of cost and utility with any other building material.

Every efficient design has its own unique characteristics. New methods of working with wood have changed the appearance of buildings. Now, a new model needs to be developed. While such a model can hardly be compatible with the current commonly held concept of 'the wooden house', this new model organically flows from and is a continuation of a centuries-old tradition of building with wood.

The intention here is to show, through illustrative examples and commentary, how working with wood in a new way can reflect a transformation in the way we build.

Berlin, October 1930 Konrad Wachsmann

WOOD AS A BUILDING MATERIAL

has been of immense importance for the building trade since time immemorial. Yet today it has almost fallen into disrepute. The explanations for this are difficult to determine. Perhaps it has to do with the fact that working with wood has only recently evolved from the craft to the manufacturing stage. Or, that one was bound too closely to traditional models, seeing wooden buildings almost romantically, as fant-

Two worlds. Both these houses were built at approximately the same time.

asies along the lines of the Swiss chalet. Or, it was assumed that cost and technical factors made wood no longer attractive as a building material: the danger of fire became an idée fixe. At best, wooden houses were considered to be provisional structures given their (supposedly) short life spans. Also, little was known concerning the insulation capacities of a well-designed wooden wall.

But when we set aside these prejudices and precisely examine the utility of wood as a building material, then we must quickly come to the conclusion that it should have, for our times, exactly as important a role to play as stone or steel. One must learn anew to utilize this material, albeit in new ways for new purposes.

One is tempted, in order to describe the contemporary state of building with wood, first to present the historical background. But one can assume that the principles of earlier wooden construction are generally known. Furthermore, many of the details of the older designs are of little relevance to today's requirements and manufacturing techniques. Only selected historical examples have been presented in order to illustrate that even the most modern designs derive from older construction methods, demonstrating that in form and concept there is a continuum between the old and the new.

To begin with, some general considerations should be mentioned that apply to wooden construction generally:
The quality of a wooden house is in no way inferior to that of a stone house. It has long been established that wood can withstand over time all normal external and internal stresses. Of course, care must be taken to ensure that the quality of the design and the materials is such that each house is able, over time, to fulfill the requirements made of it – and to amortize itself.
Governmental standards have been established for wooden structures that now facilitate their financing. These quality standards (identified by the marking DIN 1990) allow the builder, the architect, financial institutions, and the building inspector to evaluate the wooden house objectively.
The danger of fire is in no way greater in wooden houses than in others, due to the fact that the interior work – flooring and ceilings, doors and windows, etc. – is identical. Practically all domestic and foreign insurance companies accept this fact; today one can insure wooden houses under the same conditions as any other type of house.
The insulation capacity of wooden walls of all types has been established through tests carried out by governmental testing agencies. Quite aside from historical experience, the values have now been established through scientific testing.
Production and assembly methods unique to building with wood have resulted in such advantages that, indeed, in many cases a wooden

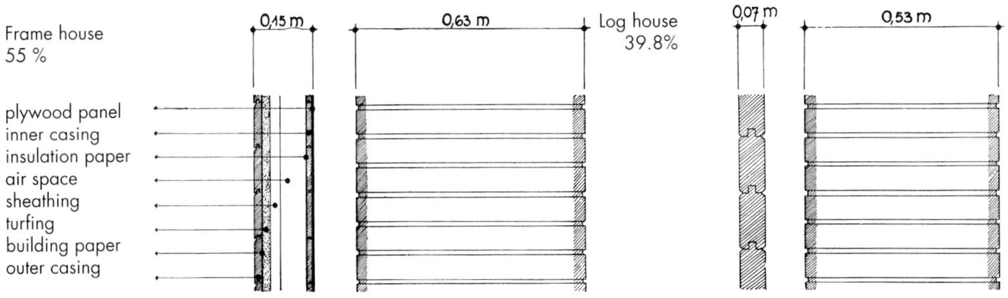

Frame house
55 %

plywood panel
inner casing
insulation paper
air space
sheathing
turfing
building paper
outer casing

0,15 m
0,63 m
Log house
39.8%
0,07 m
0,53 m

house is particularly appropriate. As most of the fabrication is done in the factory, where materials can also be stocked, work can proceed even outside the normal building seasons. Premium wages need, therefore, not be paid for out-of-season work. Furthermore, the actual construction time is considerably less than is required for houses using other materials, thereby reducing construction costs and allowing for quicker amortization.

The assembly process itself offers enormous advantages, particularly due to the fact that masonry work is only required for the foundation: the rest of the construction can be carried out in completely dry conditions

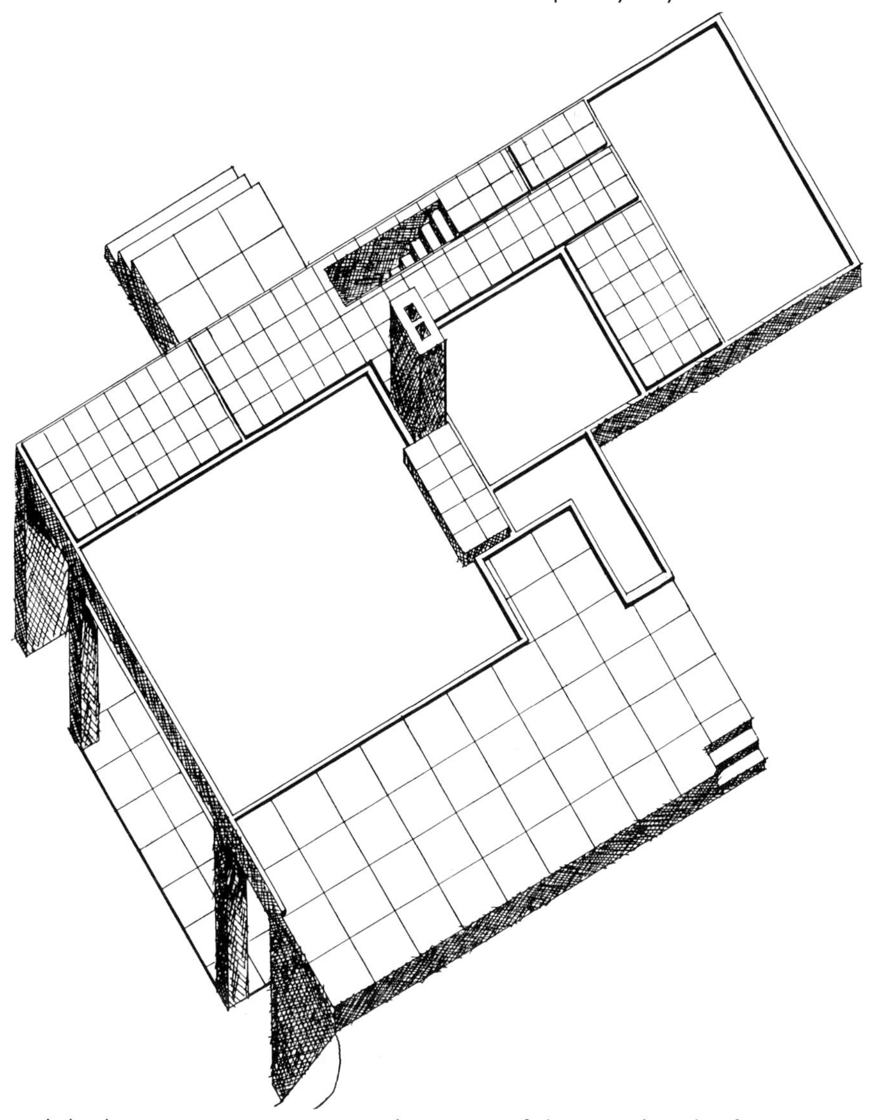

The foundation of Professor Einstein's country home. Clearly visible is the raised sill for the floor joists.16

While the various sections are being pre-fabricated in the factory, the foundation can be laid and will be completely set by the time assembly begins. This, of course, results in considerable time-saving as fabrication can take place concurrently with the laying of the foundation. At the construction site itself it is only necessary to join the pre-fabricated elements.

The house can be immediately occupied: a so-called drying-out period is no longer necessary. The house can also be assembled during the coldest weather. Frost no longer poses a problem.

The total cost of a house is generally calculated on the basis of cubic meters. In that the thickness of the walls of a wooden house is considerably less than that of a house with masonry walls (yet with the same insulation effect), one achieves a lower cubic meter total while retaining the same usable space.

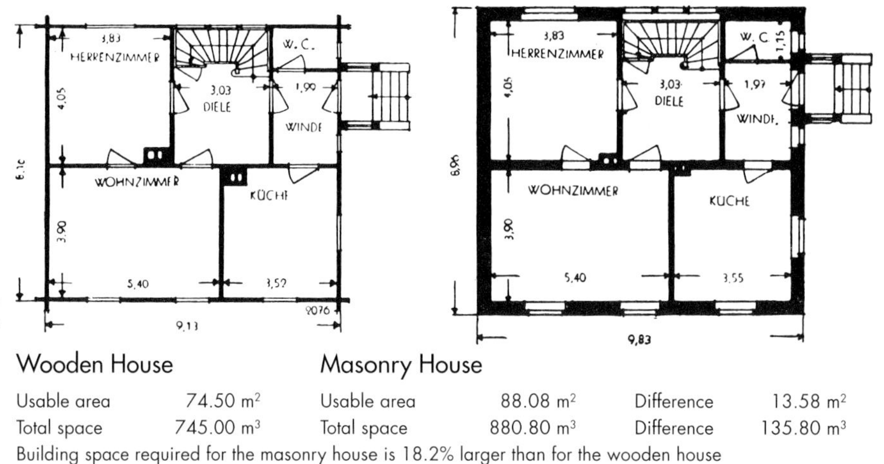

Diagram comparing usable interior space of a wooden house and a masonry house

Wooden House		Masonry House			
Usable area	74.50 m²	Usable area	88.08 m²	Difference	13.58 m²
Total space	745.00 m³	Total space	880.80 m³	Difference	135.80 m³

Building space required for the masonry house is 18.2% larger than for the wooden house

In addition, many more built-in features are possible with wood; additional space is thereby laid free since the need for space-consuming cabinets is eliminated – an indirect space advantage.

The installation of the plumbing system can be accomplished much more easily in that the space between the framing members in walls, floors, and roofs can be utilized.

But above all, the particular joy of a well-designed wooden house lies in the pleasure of rediscovering the use of wood. The natural, living material and the lightness and elegance of the building are part and parcel of the value of a house built of wood.

The combination of age-old methods with today's commercial, economic, and technical demands plus the need to meet contemporary requirements has led to the emergence of three basic building methods. These are:

1. THE ON-SITE WOOD FRAME METHOD
2. THE PANEL METHOD
3. THE LOG HOUSE METHOD

In order to be able to select the proper method for each specific project, it is essential that one understands exactly the basic differences between them.

THE ON-SITE WOOD FRAME METHOD

is not without reason the preferred method today. The development of this method has enabled the wooden house to become truly an industrial product. This method has made possible above all the planning of the contemporary wooden house so that it meets fully our requirements for economical and efficient construction, the application of technical advances, and the efficient use of space. It appears that this method provides above all others the greatest potential for technical and cost-factor improvements.

Traditional wood frame construction is characterized chiefly by the fact that horizontal and vertical members are joined using carpentering techniques. This means that joining is accomplished using tongues and grooves, dovetailing, etc., with the result that the members at these points are very much weakened. It has been necessary, therefore, that the members be of such dimensions to ensure sufficient strength at these points. But a practical and cost-effective use of this method can only be achieved if one can reduce the dimensions of the materials used to the

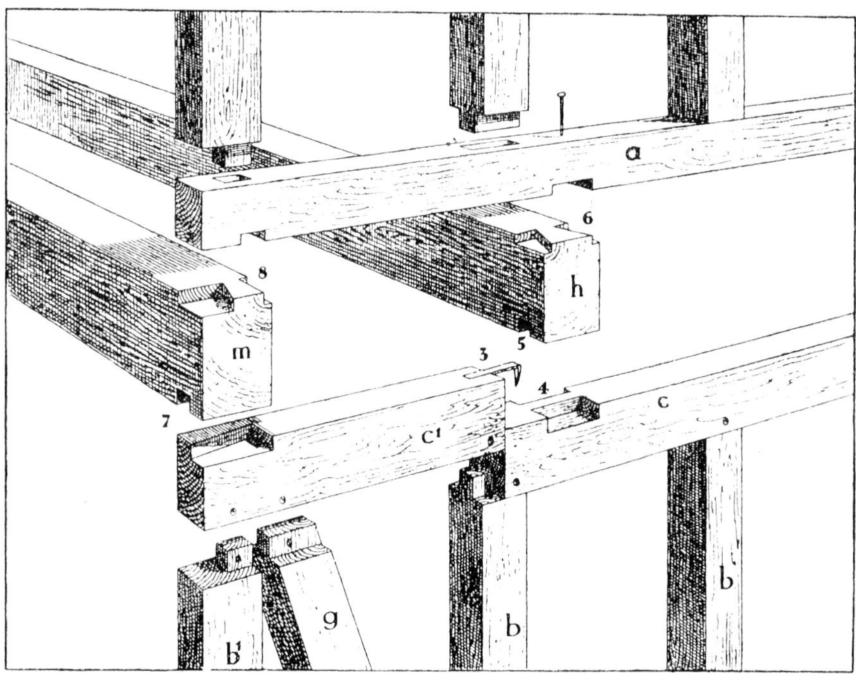

The most common methods of joining using carpentering techniques. From Büning, *Bauanatomie*

bare minimum; and if one chooses a method of joining together prefabricated framing elements that allows for simple assembly by untrained workers. This has led to the current wood frame method that differs in significant ways from older methods.

In earlier times, design dimensions were largely a matter of guess-work. A large amount of additional dimensioning was required to provide a margin of safety as stress factors could not be correctly calculated. Today, however, all critical factors can be precisely calculated, allowing from the very start for a totally different and more efficient utilization of materials. This has been true for some time in the construction of industrial structures, halls, and so on. It is now time to apply this know-

The evolution of structural strength. A wooden pillar with supporting column. The Markthalle in Nuernberg, 15th century. On the right, a modern girder construction

how to the construction of wooden houses.
Wooden houses were first constructed in this way in America. There, a new method of joining was developed that did not weaken the material at critical points. Frames and studding and all other parts of the wood frame are fastened simply by nailing. In that the wall and ceiling boards are nailed directly to the studding, beams, and top and bottom plates, the structure becomes sufficiently stable to meet normal requirements. Above all, this method is highly cost-effective.
Granted, this method does not fit in our concept of solidness and durability: by using this method the Americans have quite consciously made qualitative concessions. But on the other hand, this method has made it possible for millions of people to own homes that would not be affordable if they were built according to European standards.

In America, the mass-production of wooden houses began approximately 150 years ago. A large industry has been developed there which has an economic importance similar to the iron and steel industries in Europe. Approximately 300,000 wooden houses are produced annually in the

An American Meeting House, built using the wood frame method in 1682. It is still in use.

United States. Out of a population of 130 million, 80 million Americans live in wooden houses. These figures give an indication of the importance and scope the production of wooden houses has achieved there.
But one cannot forget that this development has been made possible because of the enormous timber resources available in America. Timber

An American farm house in Georgia. The simple design of the individual units is seen here to be particularly appropriate.

there is so cheap that even top quality lumber can be selected and still be competitive with other building material. For this reason, the houses in the suburbs of large cities and smaller towns in America are built almost entirely of wood.

Aside from some specialized design types, one can distinguish between the following three American construction methods:

the balloon frame construction
the braced frame construction
the western frame construction

The basic principle for each of these methods is the same: they differ only in their quality, which is progressively better from the balloon to the Western.

The simplest and therefore the most widespread method is the balloon frame construction. Using this method, the wall studs are continuous, extending through both levels. In the other methods, a platform type of construction is used whereby the wall framing is one storey high, in-

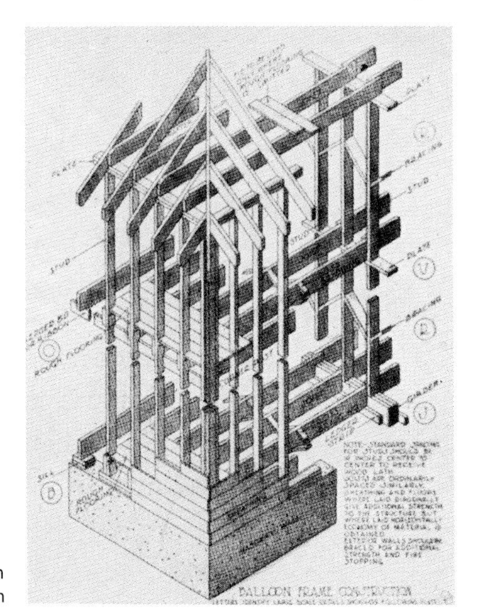

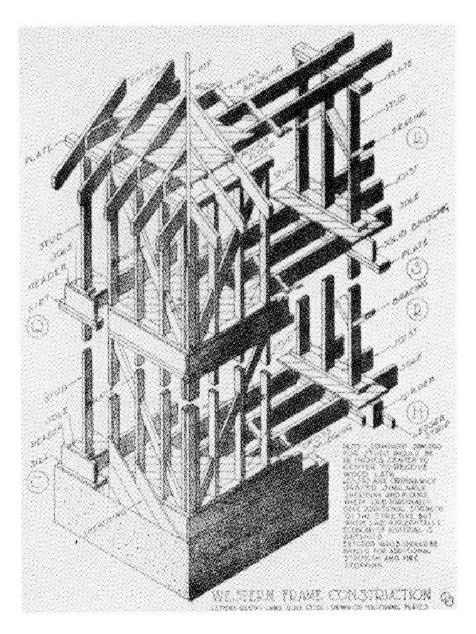

The balloon frame construction
The western frame construction

terrupted at each level by a floor frame.

The Western frame method also differs in that the boards on the wall and floor frames are laid diagonally. This provides the simplest way of dealing with shrinkage and instability and makes additional reinforcement of the frame unnecessary.

The basic principle in the construction of the load-bearing portion of the balloon frame is as follows: wall studs are spaced at 16-inch intervals and extend, in the case of multi-storey structures, continuously through all levels. Onto these, rafters are nailed and braced against each other to maintain stability. In that nailing alone is not sufficient to secure the structure, a top-plate is constructed onto which the rafters are attached.

The rafters are so placed that they attach to the studs, resulting in a kind of ribbed effect.

Throughout the design, large members are used as infrequently as possible, but where they are needed they are made up of joined softwood lumber.

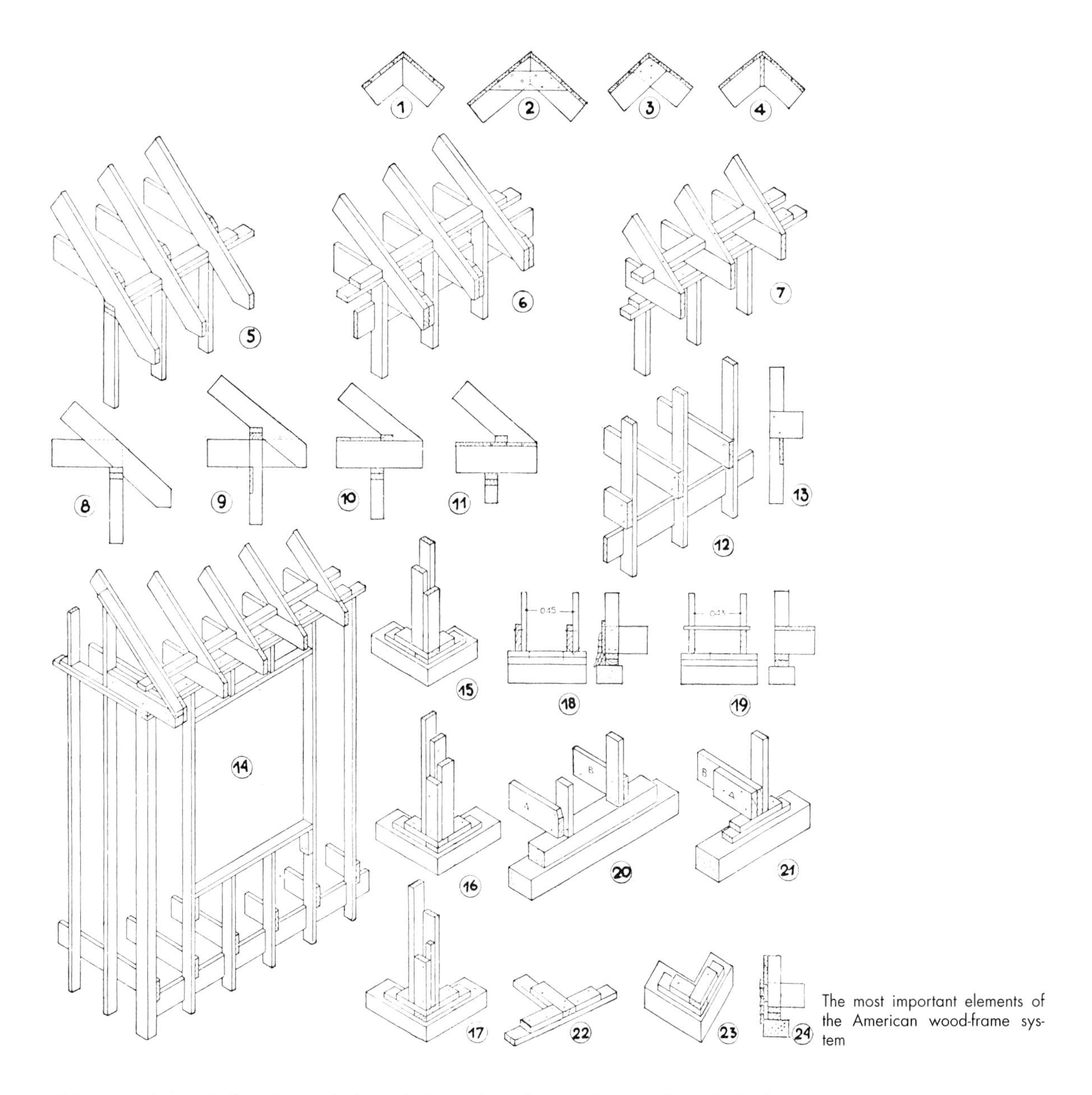

The most important elements of the American wood-frame system

Ridge joints: 1. Mitred rafters. 2. Mitred rafters with clamp. 3. Mafters joined. 4. Mafters with ridge beam. **Eave joints:** 5. Rafters attached to the top plate. 6. Rafters attached to the top plate and joists. 7. Rafters resting on a double plate. 8. Cross-section of #5. 9. Cross-section of #6. 10. Cross-section of # 7. 11. Rafters placed on the top plate. 12. Joints with joist assembly. 13. Cross-section of #12. 14. Corner assembly and other elements. 15.–17. Various corner posts. 18. Posts with stripping nailed on. 19. Posts are often nailed to the flooring. 20. Sill plate is a solid block (beam A is somewhat too short but will be used). 21. Sill plate on two solid blocks (an irregular beam is constructed by nailing A and B together). 22. A brace. 23. A corner brace. 24. Filling at the joint

All the lumber is standardized. Only through a strict adherence to standardization is it possible for the American wood-building industry to build at such low cost. Standard beam measures are:

4.5 x 17 cm
4.5 x 20 cm
5.5 x 15 cm
5.5 x 20 cm
5.5 x 23 cm The following are standard stud dimensions: 3.5 x 17 cm
4.5 x 15 cm
4.5 x 17 cm. Normal lengths are 6.9 and 9 m.

Standard-sized lumber can always be delivered quickly and cheaply. When special or custom sizes are required, however, they are expensive and require a long delivery time. Therefore, it is only rarely that special orders are made. The American typically buys from a catalogue in which a fixed price is stated – just as for an automobile or any other industrial product.

The framework of a typical American wooden construction on the second day of assembly. From Neutra: *Amerika*

The ease with which the framework can be assembled has become a critical factor in America because of the scarcity there of skilled workers. These houses can be assembled under the direction of a foreman by unskilled labor. It is possible for three to five workers to assemble and fix the elements of the normal wooden house and to apply the outer casing in approximately four days. In Germany, one would need three times as long a construction period.

The building materials are delivered to the construction site carefully packed, sorted, and numbered. All necessary tools, nails, paint, etc., are also provided. Using such exactly-machined materials anyone can, without assistance, assemble and construct his own house. The exterior siding can be done in a variety of ways, of which the simplest is the bevelled system although it is infrequently used. More common is the tongue-and-groove type in which the siding is either attached directly to the studding or to the sheathing. Plaster is also frequently used, utilizing a metal grid to separate the sheathing from the plaster. The outer sheathing forms a thin skin, separate from the wall. The outer forms a thin skin, separate from the wall. There is of course a variety of ways in which the plaster can be applied, but this need not further concern us here as the basic principles of wooden house construction are not affected.
The final stage in the construction of the pre-fabricated house consists of decoration: the addition of columns, pillars, etc. These elements are, of course, also mass-produced in America and are available in any of a great variety of types and sizes.

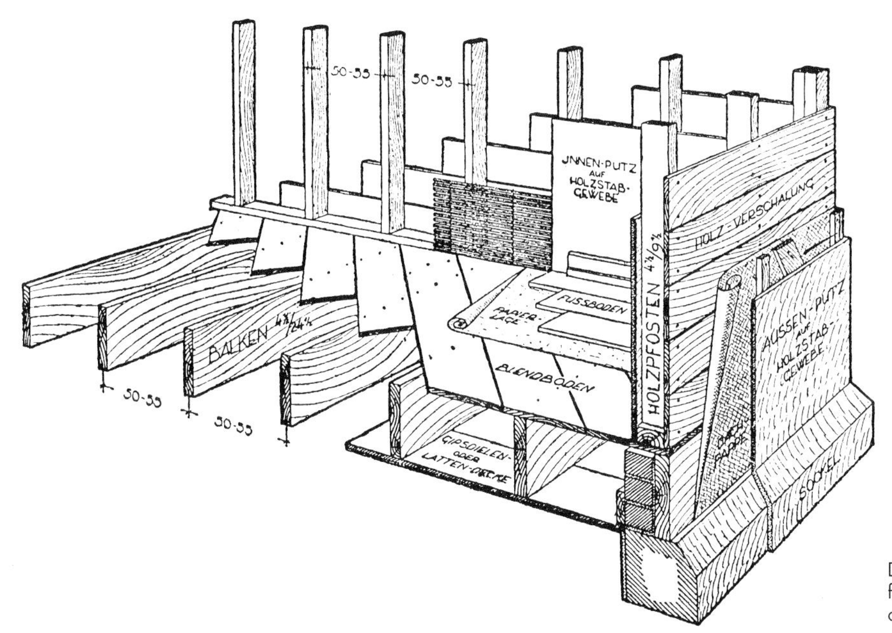

Diagram of an American wood frame construction showing the outer and inner casing

Here, we have long since overcome this tendency to excessive decoration. It has been recognized that the essential charm and beauty of a building lies in its logical and coherent design. If the architect and the builder utilize to the fullest extent possible the resources that modern technology has made available, the results – because of the clarity and simplicity of their work – will be convincing without the addition of unnecessary elements.

The introduction of steel collars that first made possible the use of the so-called dowelled beams and load-bearing trusses, has made possible the spanning of large areas. Load-bearing members can be secured at a minimum number of supporting points. Large window openings no longer present problems. This style of building with wood has created a new type of house with its own unique characteristics. Limitations of older methods have been overcome. And all the flexibility of design and construction inherent in other building materials – steel, concrete, etc. – is also possible using wood.

Combining the American practice of using the minimum necessary dimensions with German quality requirements, new methods have been developed which in no way sacrifice durability and stability to the simplicity of their design and construction.

Housing development Leupnitz-Neuostra near Dresden. Framework of a standardized German wood frame construction. The diagonal boards are temporary scaffolding.

We will here ignore the various applications of the wood frame construction method, familiar in Germany, that are extremely complex, focusing our attention instead on the simplest and most widespread method; that is, the method derived from the balloon frame construction.

For the vertical posts, 5 x 10 cm studs are placed at intervals of 50 to 70 cm. Window and door frames are 8 x 10 cm, corner posts 10 x 12 cm. One differentiates between load- and non-load-bearing walls, the latter being smaller.

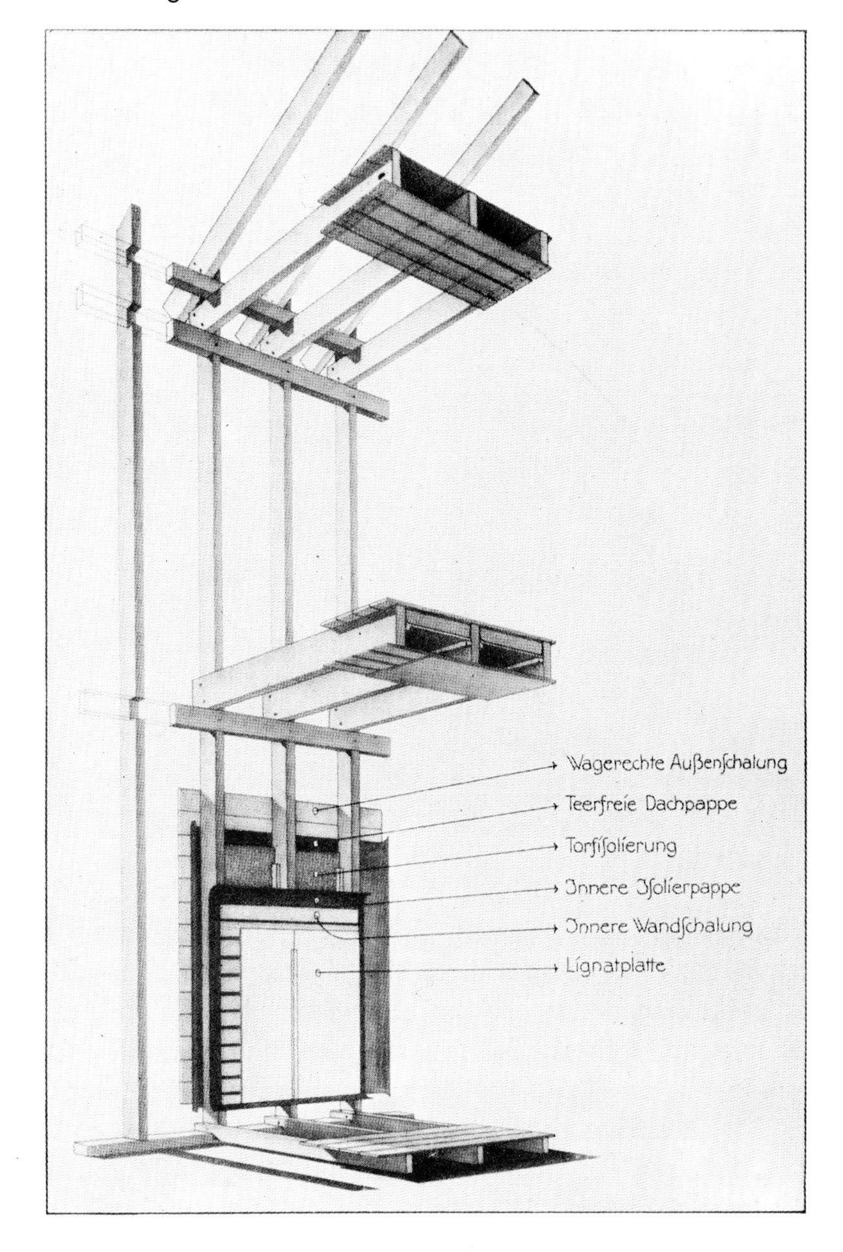

horizontal casing
tar-free roofing felt
turf insulation material
inner insulation material
interior sheating
hardboard panel

Schematic presentation of the contemporary German wood frame construction

Diagonal beams are placed at the corners, bracing the wall frame. A diagonal undercasing is rarely used because of the considerable joining involved. As in the American methods, fastening is done entirely by nailing. All the lumber is, of course, pre-cut and sorted at the factory.

To make the structure more stable, one can for example provide notched rafters for attachment to wall plates, making unnecessary wall posts extending through both storeys. A platform is placed on all sides onto which are fastened the floor beams. Along the top plate the second storey is raised in the same way as the ground floor. This method is in many respects similar to the old wood frame method.

Children's home in Spremberg. Framework showing floor beams attached to a platform. Studding is one-storey

In this type of construction, the bottom plate is lined with building paper and treated to protect against moisture. Wall studs are erected with a top plate. The framing is secured by attaching ceiling and roof beams. When the frame has been erected, the exterior, with the exception of door and window openings, is covered with sheathing paper onto which the outer casing is attached. 2-cm-thick insulation material is placed between the wall studs onto the sheathing. On the interior surface of the studding 13- to 18-mm-thick boarding is attached onto which a plywood or other type of interior finish is secured. The air space between the wall studs (8 cm) provides for excellent insulation.

In Germany, attempts have been made to develop a way to apply light-weight stone to the wood-frame as filler and surfacing. The binding of such a material to the wood frame presents considerable technical difficulties. The almost unavoidable cracks that develop over time on the stone surfaces have given this type of construction a bad name. Professor Schmitthenner in Stuttgart has devoted special attention to this problem and has experimented with ways to factory-produce the sections. He has developed a kind of frame system which allows for the joining of the wood and surface sections. All elements are standardized, including the filler sections.

Various stages of construction of a wood frame and stone-surface house, being built following Professor Schmitthenner's method. Bottom-right picture shows construction on the fourth day of work.

The vertical supports consist of split boards, 5 x 10 cm, and have been pre-assembled in the factory into 55-cm-wide templates. By screwing together these frames, wall posts are formed of 10 x 10 cm. Surfaces are then plastered with 12-cm-thick light-weight stone material, covering the exterior surface and separated from the wood frame by a 2-cm-wide air space. By separating the wood from the plaster, the danger of rot in the wood and cracks in the stone surfacing is practically eliminated.

The Frank method is the wooden house construction method with the most fully elaborated engineering. All dimensions are exactly calculated and all joints are bolted. Particular attention is given to the construction of the joints. All materials have been pre-bored on templates. Filler material, as in concrete construction, is pressed into removable forms.

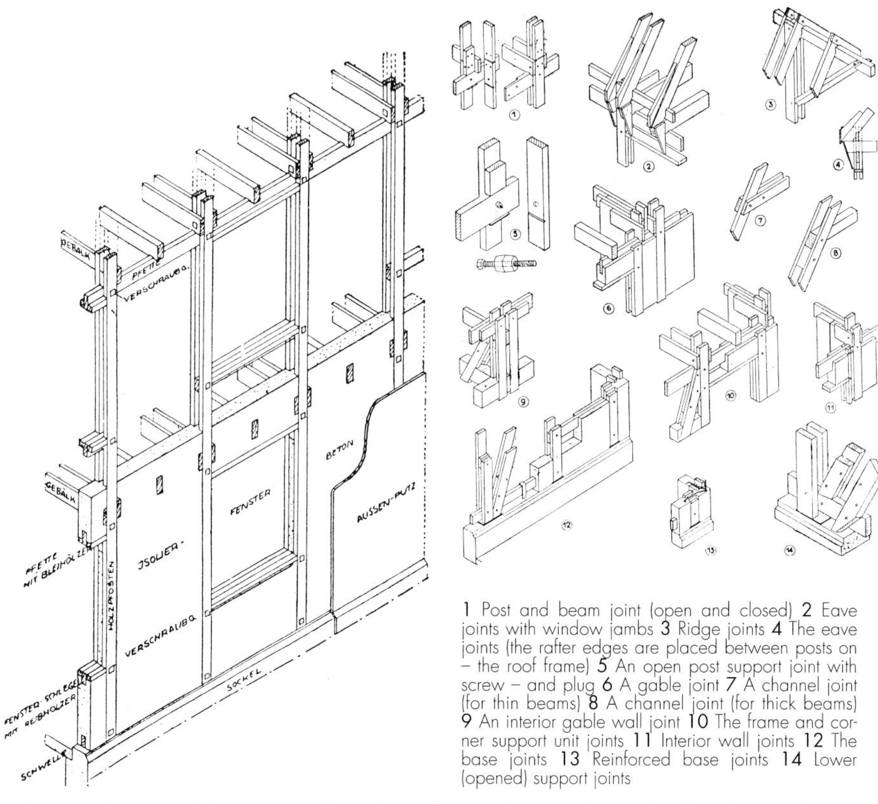

Diagram showing the patented Frank method. The most important construction details are illustrated.

1 Post and beam joint (open and closed) 2 Eave joints with window jambs 3 Ridge joints 4 The eave joints (the rafter edges are placed between posts on – the roof frame) 5 An open post support joint with screw – and plug 6 A gable joint 7 A channel joint (for thin beams) 8 A channel joint (for thick beams) 9 An interior gable wall joint 10 The frame and corner support unit joints 11 Interior wall joints 12 The base joints 13 Reinforced base joints 14 Lower (opened) support joints

Although it would appear that the relatively good results that have been achieved justify the use of the wood frame and plaster method, it should nevertheless be noted that these houses no longer exhibit the purity and simplicity of the house built entirely of wood. Above all, the advantages of a dry construction are sacrificed. Even though one has found ways to protect wood in the immediate vicinity of water, the mere fact that moisture must be introduced into a wooden house should give one pause. Also not to be discounted is the fact that, aside from the interior work, a house entirely of wood requires only one type of construction worker, not two.
Summing up, we can state that the on-site wood frame building with wooden sheathing is capable of meeting all construction requirements. Through the utilization of the wall as a load-bearing framework which can be encased with a variety of materials, a type of construction is -

achieved that meets the most up-to-date requirements – and one that has significant cost advantages.
In that we in Germany do not have the same, largely uniform conditions that prevail in America, and because we here must take into account a great variety of cultural – even climatic – factors, it has been difficult until now to develop the German wooden house along the

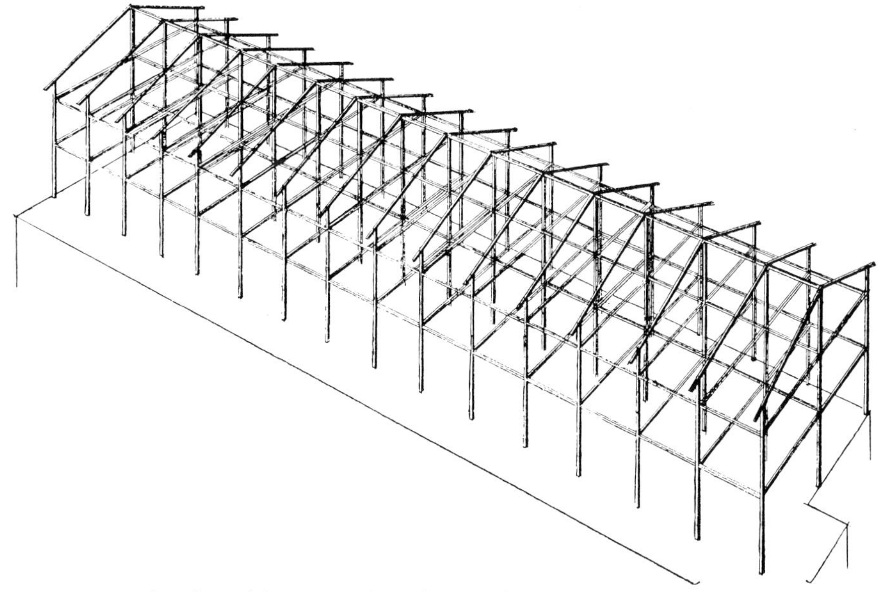

Youth hostel on the Spindlerpass. Load-bearing elements

same standardized lines as has been done in America. But we can see that now serious efforts are being made to develop a fresh common denominator for the requirements we place on our wooden constructions – largely unencumbered by sentimentality. There is every indication that the German wood building industry has absorbed much from the American experience.

Street view. Wooden house development, Leupnitz-Neuostra, near Dresden

THE PANEL METHOD

allows for the fullest possible standardization in the construction of the wooden house. All the building parts in this type of construction are pre-fabricated, including the wall, window, door, ceiling, floor, and roof panels. By standardizing interior arrangements and dimensions, it is possible to produce all of the individual panels in large quantities. According to individual requirements, all of the panels necessary for every type of design can be ordered from pre-stocked supplies.

roof panel

attached rafter

ceiling panel with insulation

panel frame with support

horizontal casing

outer insulation material

filler insulation material

turf insulation material

inner insulation material

interior sheathing

hardboard panel

floor panel

Diagram showing parts used in the panel method

The interior part of the wall panel is constructed following the same general principles as apply in the case of a wood frame house. On an approximately 3-cm-wide bottom plate a horizontal tongue-and-grooved casing is fastened. A 2-cm-thick insulation panel is nailed in and then a 2-cm-wide reinforcement panel, onto which a dry wall is attached that serves as a mounting for plywood or for any other type of interior paneling.

Window paneling is constructed as are the wall panels, except that window or door panels are pre-framed and ready for installation. The entire assembly can be transported as a unit.

Pre-fabricated wall panels for a large hotel in South America, shown in a warehouse in the harbor of Curaçao. The house was delivered in 1929 by a German company.

During assembly a bottom plate is attached as usual to the foundation, well-protected against moisture, and grooved so that the wall frame can be joined.The panels are designed to overlap so that a completely tight fit is achieved.

Laying of the bottom plate on the foundation. Hotel in South America

For additional stability, battens may be attached on the exterior and interior of panel joints, giving this type of house, particularly on the

Erection of the ground floor wall panels and assembly of the ceiling beams

exterior, a distinctive appearance. But it should be noted that the use of pattens, particularly if the design has been carefully thought through, is not always necessary.

The wall panels are fastened with metal brackets which have been fitted in the frames. When all the wall panels have been erected, the ceiling joists, also fitted with metal brackets, are installed and then the rafters are placed.

The ground floor has been erected. The panels have been painted in the factory and the window units have been preassembled.

Finally, following the same procedure, the roof, ceiling, and floor panels are installed. All paneling must be so dimensioned that it can be easily handled by two workers – only in this way is ease of assembly made possible.

View showing the rafters onto which the roof panels will be laid that have been covered with insulation material.

As all individual panels are totally pre-fabricated and finished, the work at the construction site is reduced to a minimum. There are many construction jobs, for example at a remote site, when this becomes a critical factor. And, this type of house can at any time be easily dismantled and reassembled at another site with no loss of material.
Of course, the use of so many individual elements is costly. The solidity of construction is also less than with rigid construction methods. Apartment houses, therefore, do not lend themselves to this method of construction. On the other hand, this method is particularly suited to

the construction of pavilion-type structures for communal uses: classrooms, hospital wards, buildings for temporary housing – all are ideal for this method. Here the flexibility and customizing inherent in this method is of particular importance. For example, communities that have quickly sprung up in industrial areas often need to have accom-

Panel method. Every element used in the construction of this school was pre-fabricated.

modations immediately available for public welfare purposes and so on. It is not always possible to meet these demands with large, substantial structures. The panel method is an ideal solution in that the elements can be quickly delivered. They can also be disassembled for use at another site should that be required.

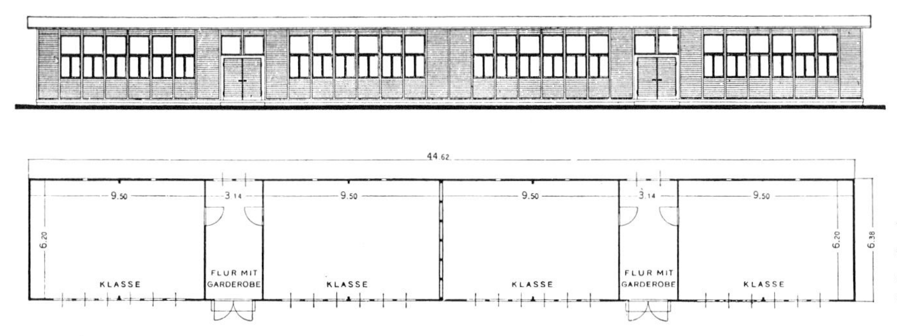

Through the use of standardized wall elements, this basic design can be constructed in any size and configuration.

It is of the utmost importance, however, that all building elements used in this method of construction be standardized; this is absolutely necessary in order to achieve cost-effectiveness.

THE LOG HOUSE METHOD

is the oldest method of constructing wooden houses and has been used since time immemorial. This system evolved in almost every country and continues to be used up until the present day. Actually, the log

A Swedish log house, built in 1758. A particularly fine example of a simple construction that, due to its few, regularly placed windows, retains the purity of its design.

house comes closest to our idea of a wooden house in that, apart from the simplicity and functionality of its design, it displays the natural wood in its purest form. And despite all refinements in design the basic

An old Swedish farm house, with gallery and outside stairs

principle of the log house remains the same as in the earliest and most primitive examples.

Today, log houses are built in the following way:

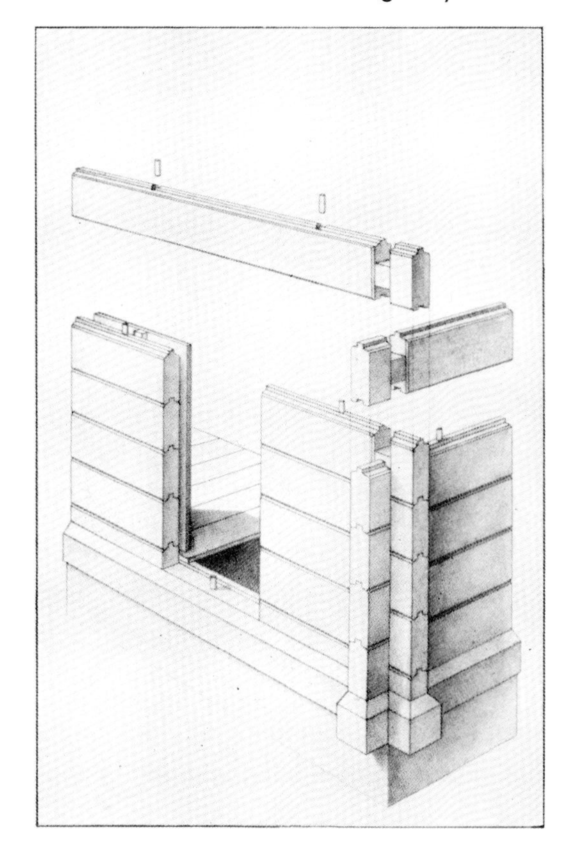

Diagram of a log plank wall

On a notched bottom frame that rests on the foundation, logs of approximately 16 x 7 cm, fitted with tongues and grooves, are laid one on top of the other.

Start of construction of a log house. The vertical joints of the logs are always staggered.

To fasten the structure as securely as possible, the log planks are fitted at certain intervals with pegs which are inserted into the prepared holes during construction.

To make sure that at the corners the upper edge of a log will intersect at the center of the log perpendicular to it, the logs are so laid that split and full logs are placed alternately.

Pegs on the top log can be clearly seen. Also visible are the grooves made at the cut of the logs to accommodate tongues.

At the corner joints, a quarter of the log is shaved off at both top and bottom.

At the corners, the logs extend by 25 to 30 cm to give them additional strength and protection at these critical points. The system of corner-joining is one of the centuries-old characteristics of this method and one which has remained unaltered to the present day. Interior walls are dove-tailed to rafters.

The completed ground floor. Floor joists can be clearly seen

At windows and doors, whose openings have been cut during assembly, the ends of the logs are fitted with grooves into which a tongue is fitted after assembly to protect against slippage. This also allows for the fitting of the door and window assemblies as this would not be possible directly onto the cut end of the log.

Floor beams are laid as in any other type of construction. It is, however, important that the wall logs, into which the floor beams are inserted, are larger than the floor beams, providing sufficient surface so that the ends of the beams are not stressed and will not split.
If one does not wish to disturb the outer surface of the wall, additional reinforcement can be easily installed on the interior side of the wall.

The ceiling beams have been laid and the upper storey can now be installed.

If, however, the log ends are allowed to protrude, a particularly pleasing effect is achieved.
The protruding logs must be fitted with a metal protective cap in order to prevent water seepage between logs and beams. As with all detailing in wood, particular care must be taken that no moisture is al-

The upper storey is completed. The protruding roof beams can be clearly seen; the gutter will be placed upon these so as not to be visible

lowed to penetrate. In the case of unusually large window openings it is of course necessary that the logs forming the frame be reinforced.

The rafters are as usual laid upon a top plate. This type of construction has a unique characteristic in that the logs are liable to shrinkage

As with all other constructions, the final stage of the framing is the installation of the roof structure.

during an initial two year period. The width of the logs – in an average storey of three meters – can shrink by as much as 10 cm. In the case of a two storey house this can be of considerable importance. This factor must of course be taken into account in the overall design of the house. All door and window framing must be so designed to allow for the weight of the logs above. The interior surfaces, whether of dry board or a similar material, must also be so constructed that they can absorb the weight of the floor timbers. This is accomplished by placing horizontal stripping behind the wall surfacing and attaching it to the ceiling. This leads to a certain "heavy" effect, but is typical of log houses. The cracks between the logs that could come about through shrinkage will generally be sealed by the settling effect of the logs themselves. What must be observed, however, is that the load must be as evenly distributed as possible to guard against both cracks and misalignments in the walls.

In order to ensure that the top-most logs of the wall structure are under sufficient pressure, it is advisable to install a sharply gabled, heavy roof structure that can function as a natural topweight.

View of a director's residence in Niesky. The previous depictions of the various construction stages are of this residence.

Flat roofs can, because of insufficient weight on the upper log layer, lead to cracking. Therefore, if a flat roof is to be installed it should have sufficient overhang to take this factor into account and protect the uppermost logs.

A log house with wide eaves, a flat roof, and an overhanging upper storey, which is a wood frame construction

It should also be pointed out that due to the considerable weight of log walls the foundation must be of corresponding strength.

The insulation capacity of log walls is outstanding. This has been proved over the years in both the cold climates of Russia and Scandinavia as well as in the tropical climates of South America.

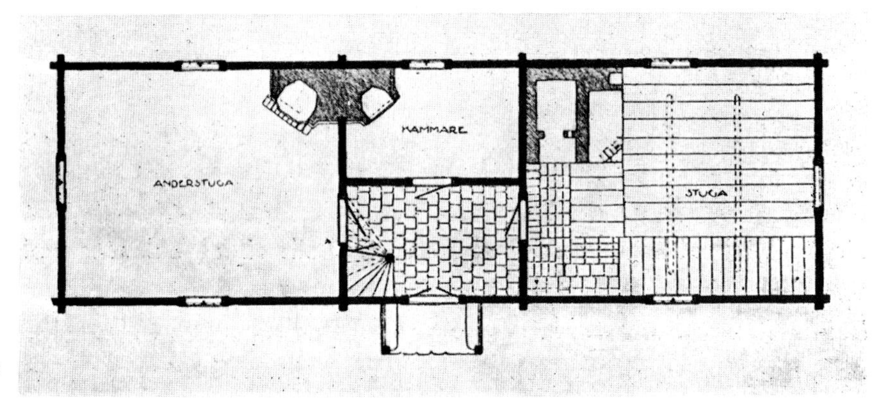

View and plan of the log house on page 30

The insulation capacity can be increased even more if insulating material is applied between the outer wall and the inner sheeting.
Piping must be installed with particular care. Here also attention must be given to the characteristic downward pressure of the logs. In that all piping must be installed with a certain flexibility, it is necessary that this work be done by experienced workers. When piping must be extended through walls, a simple hole is not sufficient; an opening must be made that allows for the free movement of the piping. Particular care must also be taken during the chimney installation and application of the surrounding roof sheeting as the chimney of course remains rigid while the roofing will settle.

The log house depicts, so to speak, a semi-rigid system. Characteristically, the walls are formed of whole logs – which means the use of a large amount of raw material. This is the main reason why the log house method has only been widely developed in those countries with large timber resources. In Germany, the log house is by and large a matter of custom-building. Mass production is hardly possible as construction, despite all technical means available, remains a matter of precision work requiring skilled carpenters.

Log house construction requires several skilled carpenters as considerable on-site work is needed.

Every log house must be test-assembled in the factory or sawmill to ensure that all parts fit snugly. Even then, additional fitting is required at the site.

But despite all its disadvantages and its high costs, it remains true that the log house is the most substantial imaginable of wooden houses, one that is almost indestructible. This is testified to by the existence of centuries-old buildings still in use all over the world.

LARGE, NON-RESIDENTIAL WOODEN BUILDINGS

do not really fall within the purview of a discussion of the wooden house. They are mentioned only briefly to suggest that wood as a building material is suited to every imaginable building requirement – that there is almost no design challenge that cannot be met with wood. Indeed, in many cases wood is preferable to other materials. As an example, the German Railway System has had all of its locomotive sheds built of wood. Exhaust fumes that would damage iron or steel are actually beneficial to wood and help to preserve it. Similar observations have been made in the case of warehouses used for the storage of chemical fertilizers. The following illustrations of wooden halls and auditoriums are only suggestive of the contemporary uses of wood in such types of constructions.

The Westphalia Hall in Dortmund. One of the largest wooden halls in the world with a span of 76 meters

The rafter and bracing system of the Westphalia Hall

The Broadcasting House in Berlin. Plaster and wood frame. Professor Heinrich Straumer, architect

The triangular truss system of the Broadcasting House, Berlin

Details of the Broadcasting House extensions. General view of the Broadcasting House

Large chemical storage warehouse near Sollstedt during construction

View of the truss system of the a chemical storage warehouse near Sollstedt

Large chemical warehouse of the Rossleben Cooperative

Large market hall in Breslau. City Building Councillor (retired) Berg, architect

The following section is intended to illustrate the preceding general discussion. A broad survey of the techniques and designs involved in building with wood should serve to illustrate the major points.

Current developments in building methods are so much in flux that it would be difficult to describe in great detail all aspects of on-site building methods. One cannot say that a wooden house should appear this way or that way. If, as has been argued here, it has been well built and if it has a clear and coherent design the goal has been achieved. Whether, for example, a wooden house should have a steep or a flat roof can never be determined generally. Each new task presents its own demands and will, therefore, always require new solutions. These, as long as they remain faithful to the materials themselves, will always have their unique attributes.

One may paint a wooden house white and certainly achieve a quite remarkable result; one may also apply lacquer so that the natural wood is shown – and to provide protection against weathering. But one may also leave the wood unfinished – particularly appropriate in the case of log houses – as weathering is in fact the best way of preserving wood. In that way the most subtle coloring is also achieved: a distinctive silver-gray.

Similar considerations apply to the interior design of the house. One may leave the horizontal flow of the timber exposed, as in a log house, or, as in a frame house, a casing can be applied. The effects that can be achieved through the use of plywood panels are well known. The use which is increasingly being made recently of special casing materials [such as Lignatplatten] and moldings gives a unique look to interior wall surfaces.

All these questions remain open, each must find its own solution. In this process everyone becomes a direct contributor to the further development of the art of building with wood. If this book has provided some inspiration along these lines then it will have accomplished its purpose.

THE ON-SITE WOOD FRAME METHOD

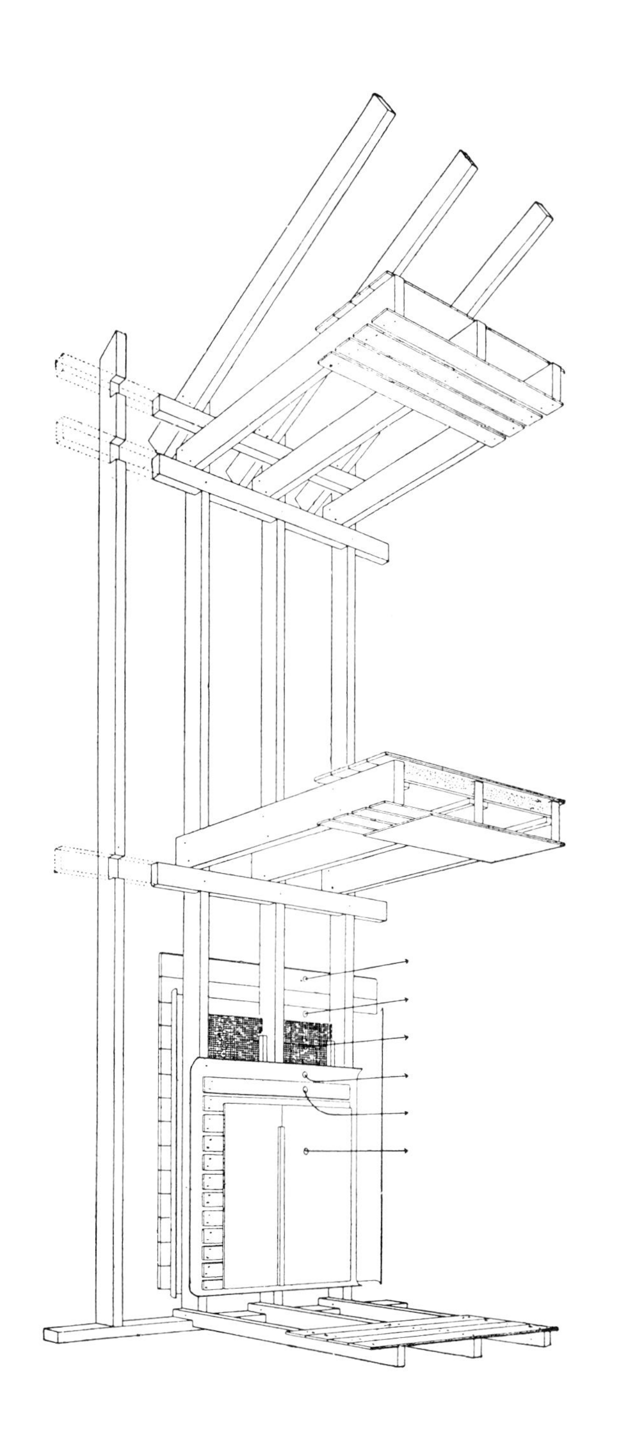

Main entrance. Taken during construction in winter

Children's Convalescent Home in Spremberg. General view with the main entrance. Konrad Wachsmann, architect

Children's Convalescent Home in Spremberg. The dormitory wing

Children's Convalescent Home in Spremberg. The south side. Large living room with sliding windows

Model of the entire structure

Children's Convalescent Home in Spremberg. View of the large living room wing, south side

Children's Convalescent Home in Spremberg. Lounge area with steps to the roof terrace

Framework and casing of this section, taken during winter construction

Children's Convalescent Home in Spremberg. The service court showing the windows of the lavatories

Model showing the entire structure except the ground-level lounge

Children's Convalescent Home in Spremberg. The roof terrace with open-air hot and cold showers. These are controlled from the inside. The flooring is covered with zinc. The gutter is in the middle. Visible in the picture is the planked grid on the surface.

Children's Convalescent Home in Spremberg. Ground floor plan

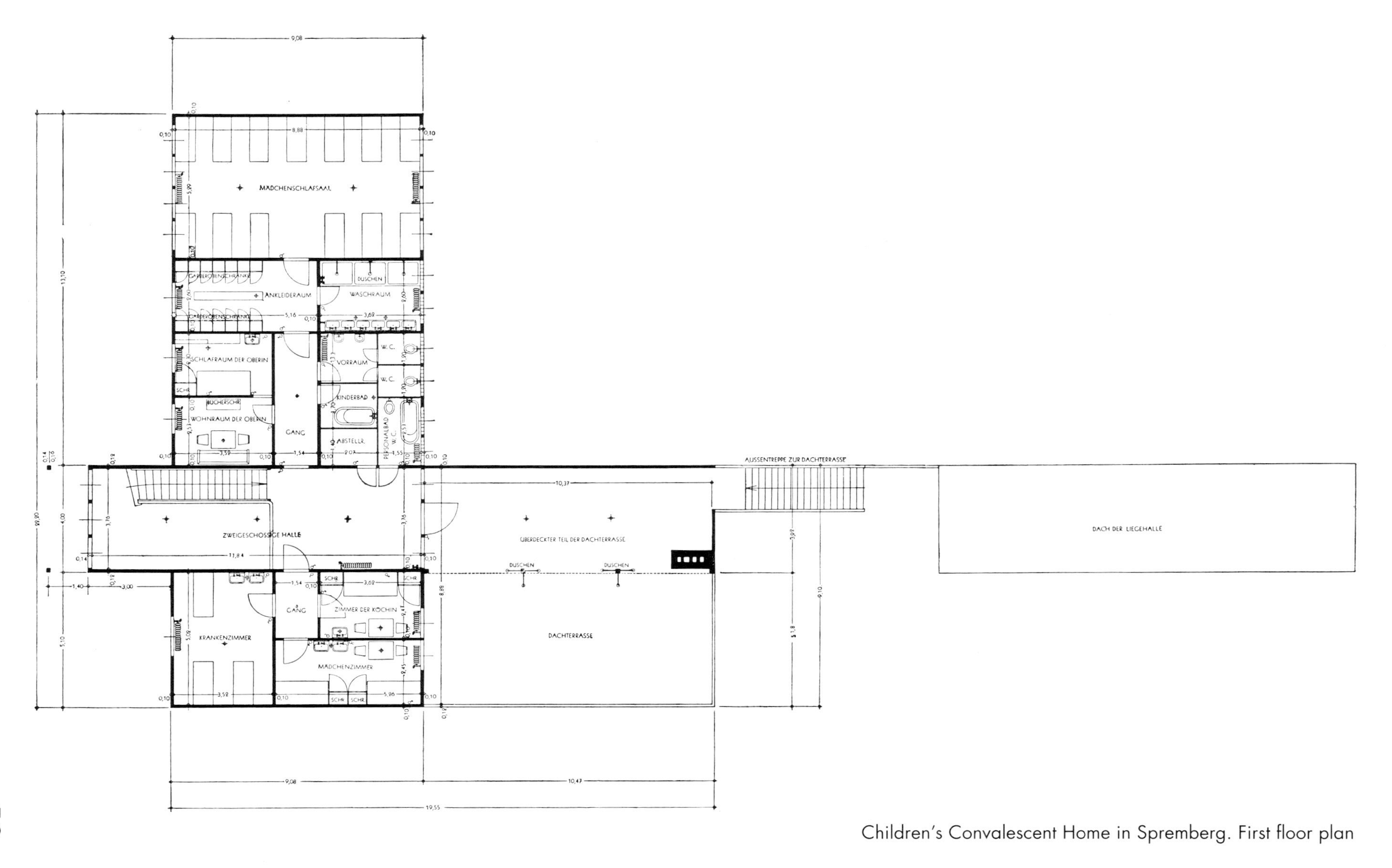

Children's Convalescent Home in Spremberg. First floor plan

Children's Convalescent Home in Spremberg. The main entrance hall and the main staircase

Children's Convalescent Home in Spremberg. View from the upper gallery toward the entrance hall

Opened hinged windows over a bench at the small side of the living room

Children's Convalescent Home in Spremberg. The large living room showing sash windows on each side

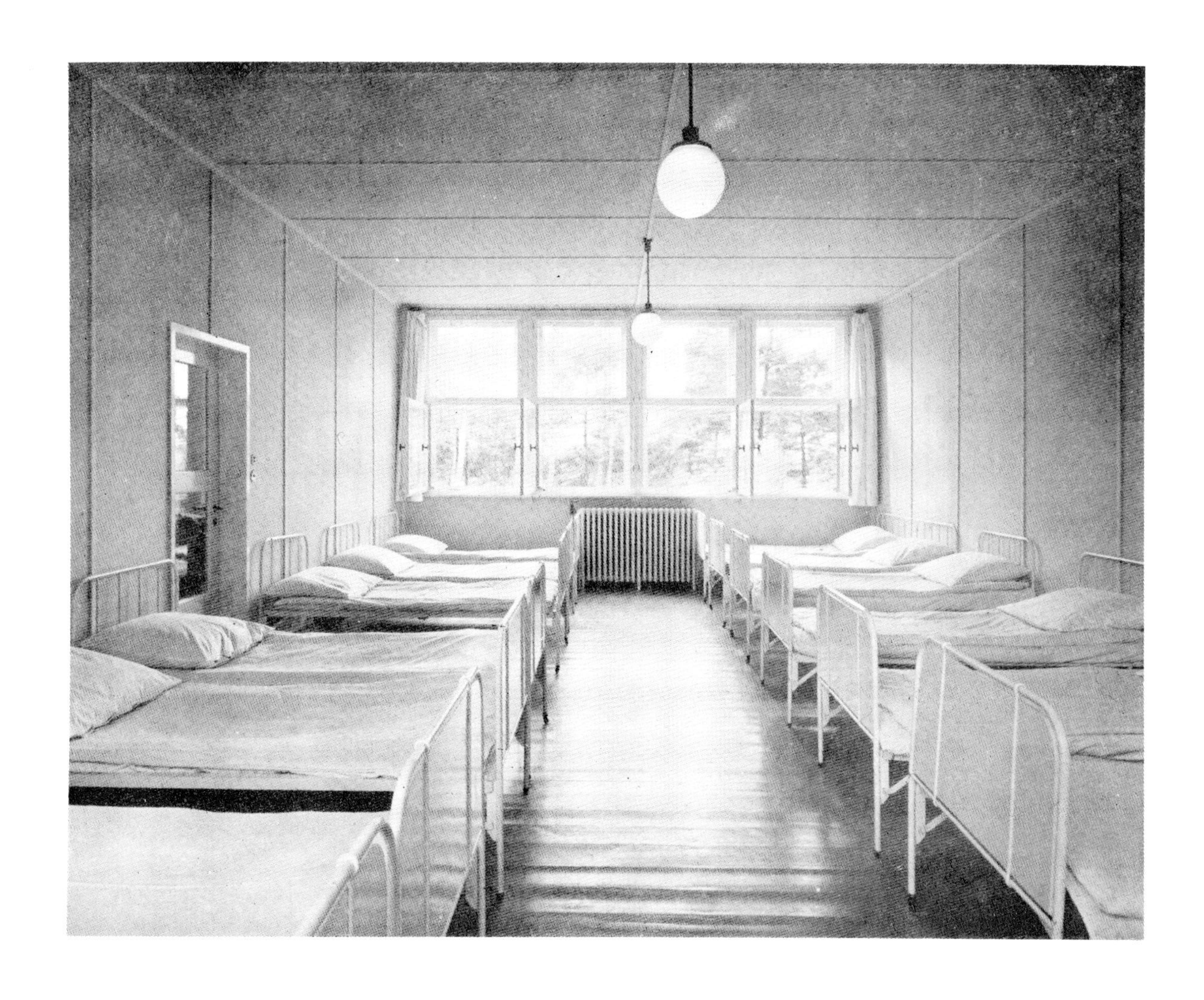

An isolation ward

Children's Convalescent Home in Spremberg. View of one of the main dormitories. Windows are at both ends of the room

Section of the kitchen with stove. The staff dining table can be extended. Terrazzo floor

Children's Convalescent Home in Spremberg. The kitchen showing built-in cabinets with glass fronts that can be raised and lowered. 6-meters-long work table under the windows with built-in sinks. Serving openings in the far wall

Dressing room adjacent to the bedrooms. Ventilated closets

Children's Convalescent Home in Spremberg. View of the children's washroom. Sinks line the wall. Hot and cold water mixture is centrally controlled by a nurse. Stone flooring

Section of a single-family rowhouse with southern exposure. View of the south side. Lüdecke, architect, Dresden

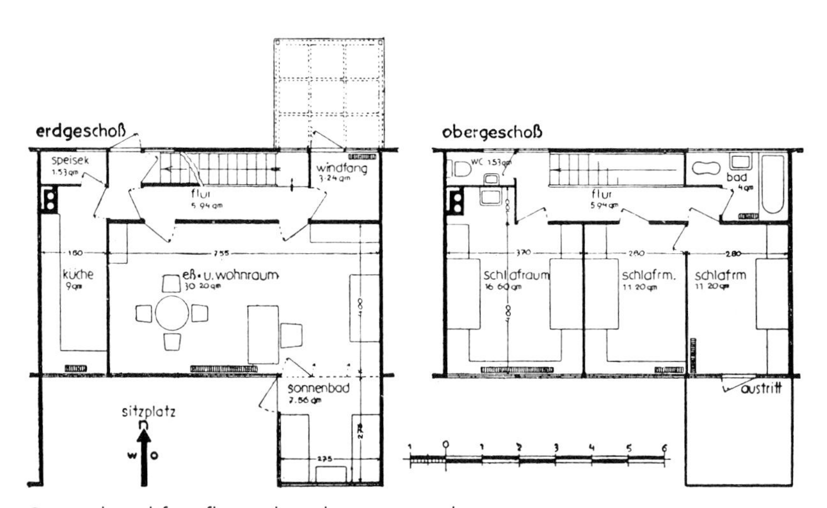

Ground and first floor plan showing southern exposure

Bedroom windows with built-in cabinets and radiators

Windows of the main living room with blinds

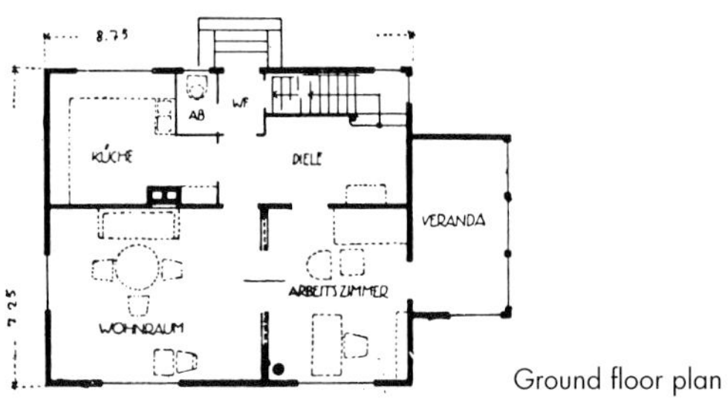

Ground floor plan

The Leupnitz-Neuostra Housing Development. Garden view of a single-family house. Eugen Schwemmle, architect, Dresden

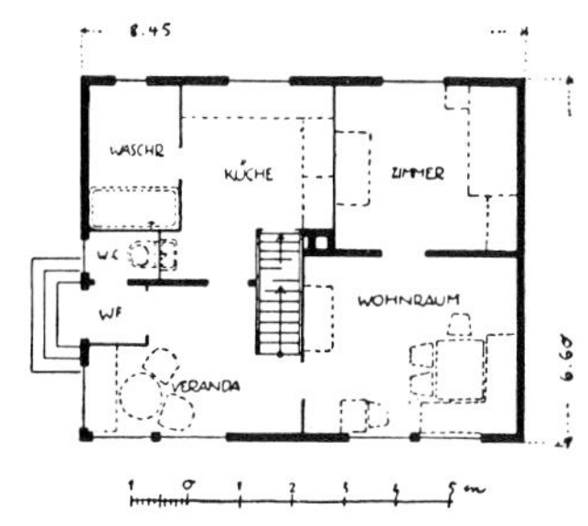

Ground floor plan

The Leupnitz-Neuostra Housing Development. Street view. Eugen Schwemmle, architect, Dresden

The Prohlis Housing Development, near Dresden. Three-family houses. Eugen Schwemmle, architect, Dresden

A view of the Leupnitz-Neuostra Housing Development. Eugen Schwemmle, architect

A view of the Leupnitz-Neuostra Housing Development. Eugen Schwemmle, architect

A view of the Leupnitz-Neuostra Housing Development. Eugen Schwemmle, architect

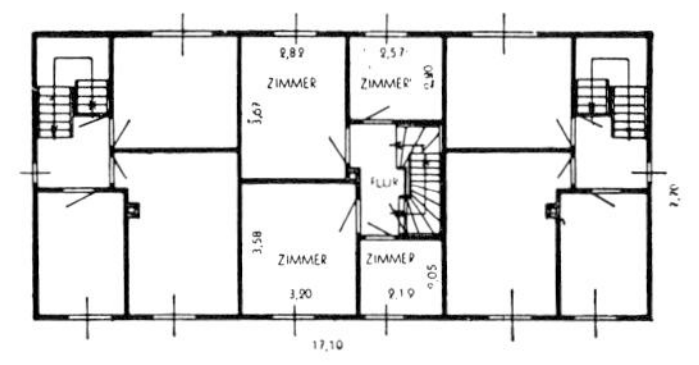

First floor plan

Scaffolding erected for painting

The Leupnitz-Neuostra Housing Development. View during the construction of three-family houses. Professor Oswin Hempel, architect, Dresden

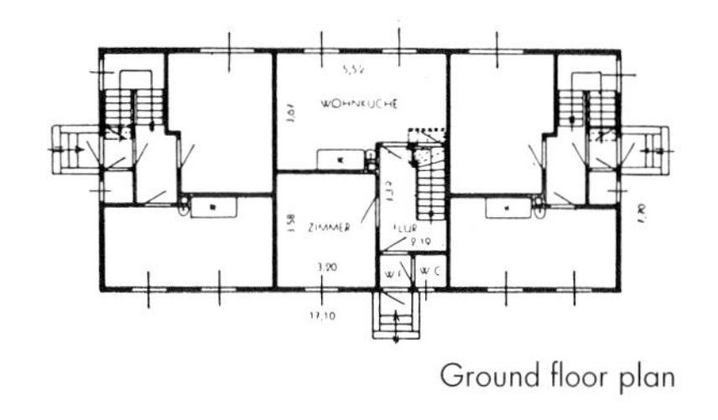

Ground floor plan

View of the completed houses

The Leupnitz-Neuostra Housing Development. In the foreground, the framework of a three-family house

The bottom of the outer staircase seen from the large roof terrace

Country home of Professor Albert Einstein, garden view. Caputh near Potsdam. Konrad Wachsmann, architect

View through the entrance door to the terrace adjoining the living room

Country home of Professor Albert Einstein. Rear view with the main entrance

Country home of Professor Albert Einstein. Recessed garden room. Above, a living room window and bannister on the roof terrace

Country home of Professor Albert Einstein. The outer staircase. Shown with the covered terrace adjoining the living room and a portion of the large roof terrace

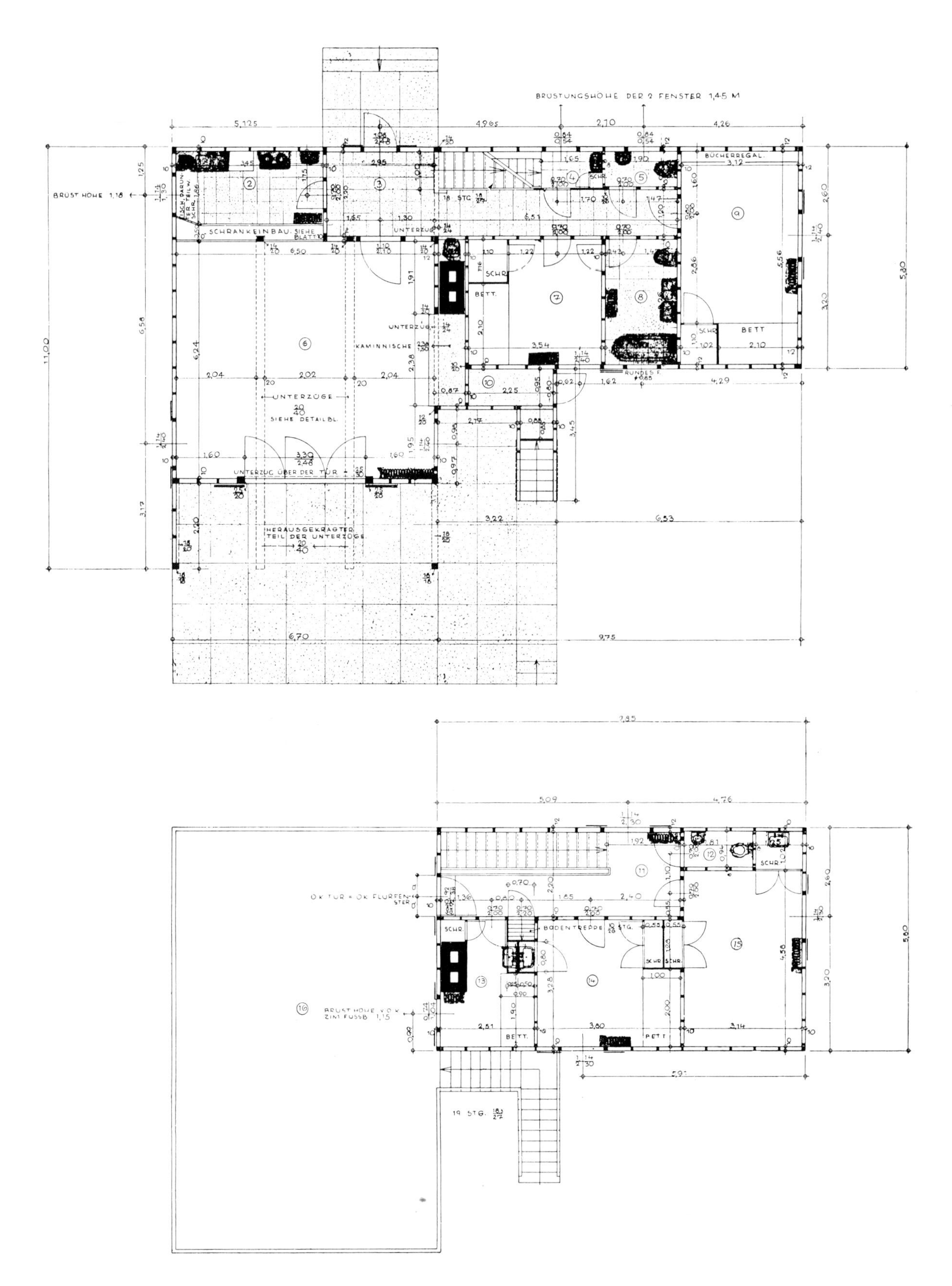

Country home of Professor Albert Finstein. Ground and first floor plans. Konrad Wachsmann, architect

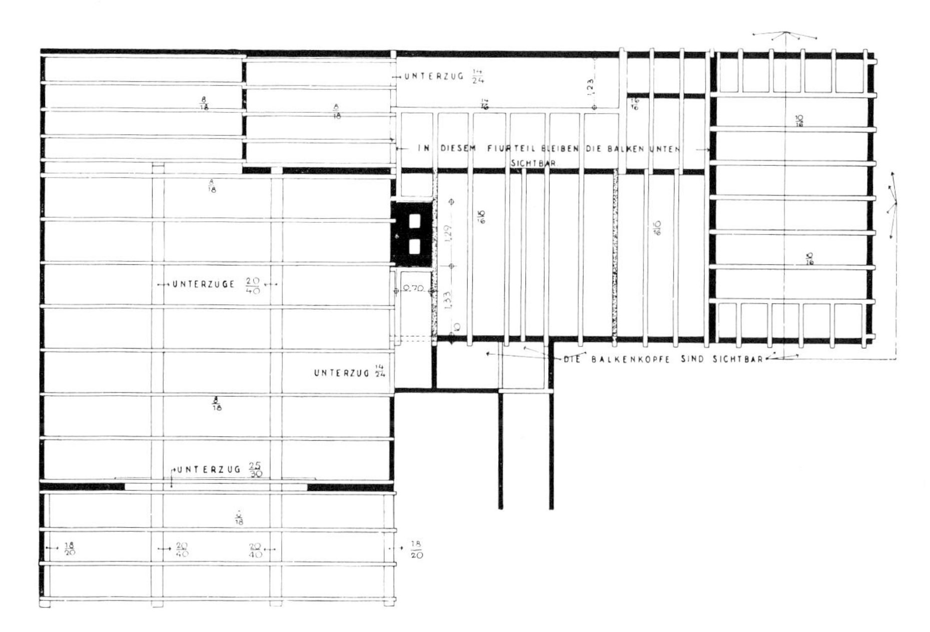

Floor beams of the ground floor showing the recessed living room

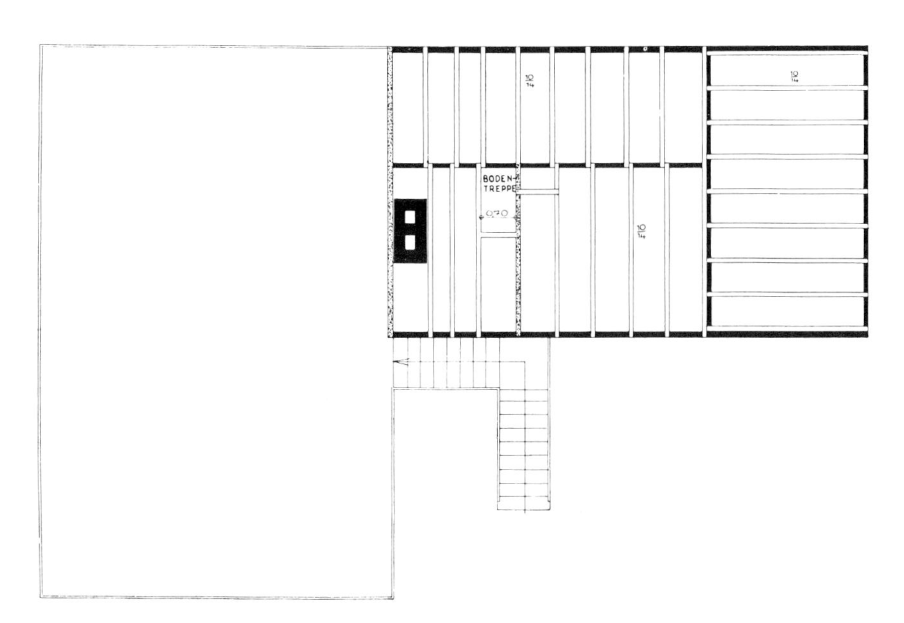

Country home of Professor Albert Einstein. The first floor beams

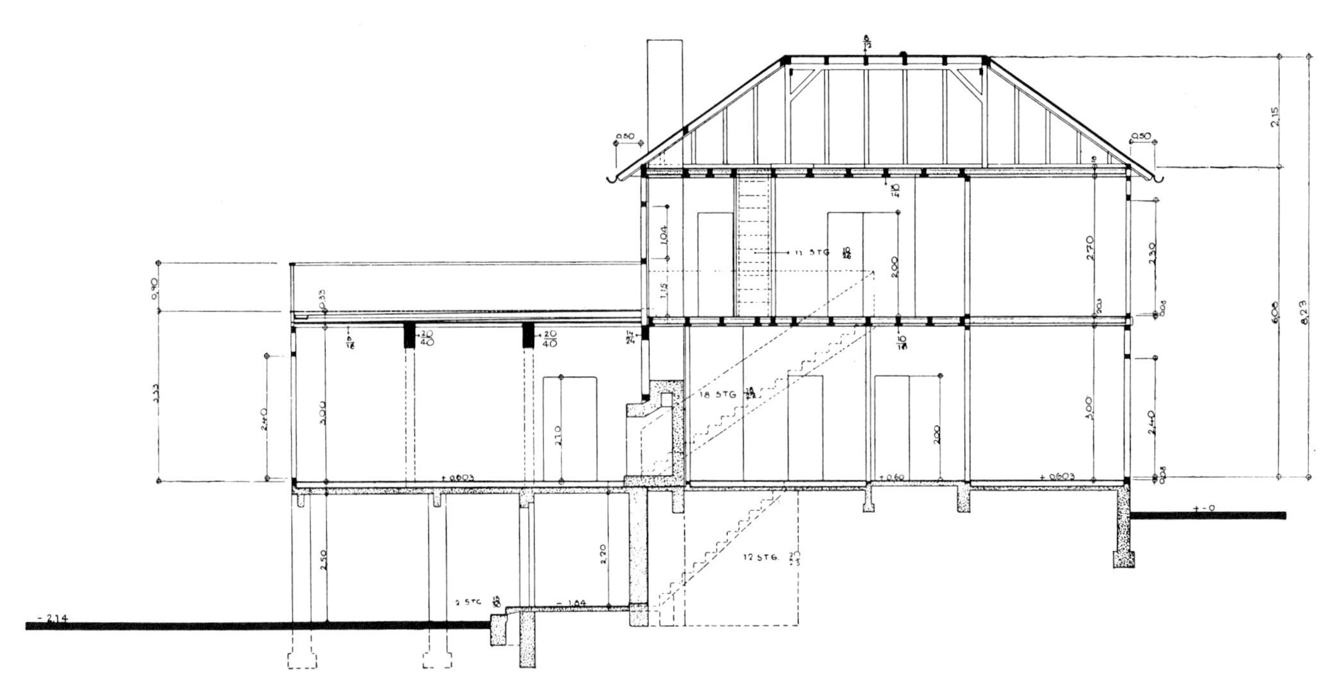

Country home of Professor Albert Einstein. Longitudinal section of the two-storey structure

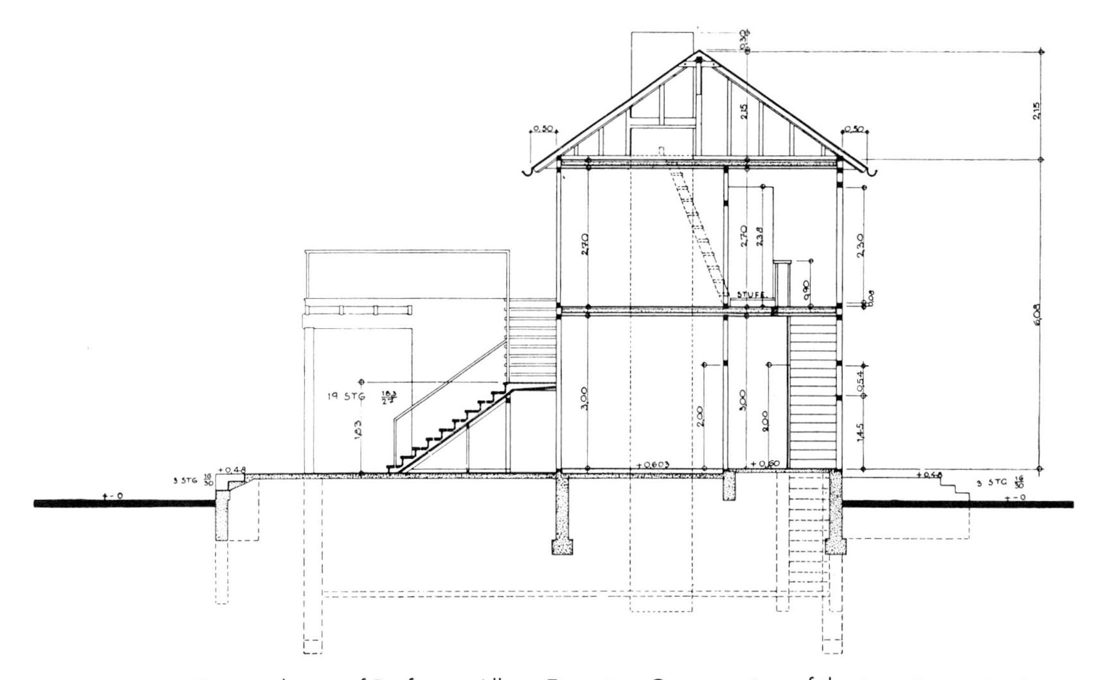

Country home of Professor Albert Einstein. Cross-section of the two-storey structure

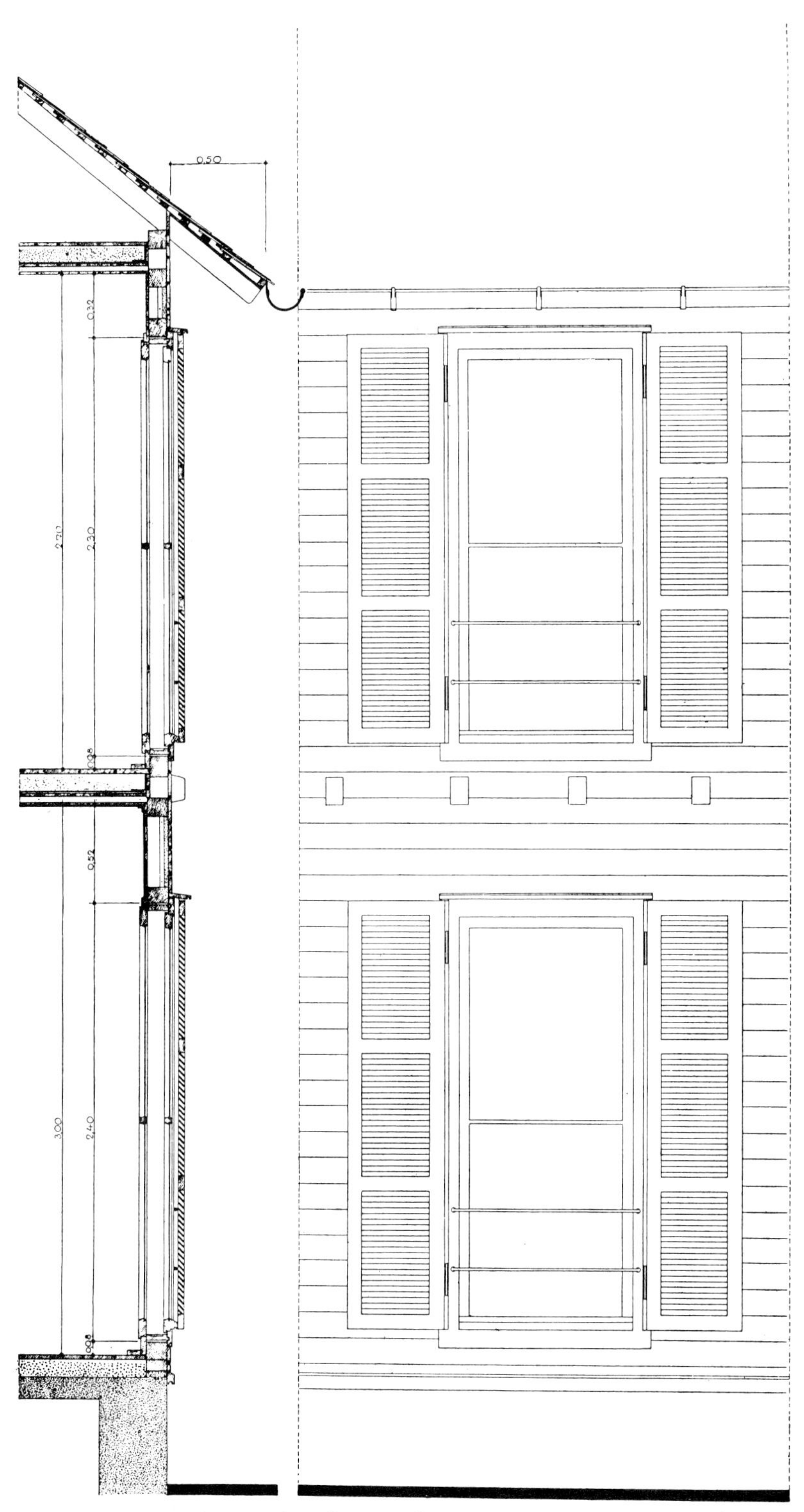

Country home of Professor Albert Einstein. Front view and cross-section of the double windows with shutters. Windows extend to the floor

Country home of Professor Albert Einstein. Open fireplace in the large living room. Portion of the wall with plywood finish

Country home of Professor Albert Einstein. Hallways with stairs and tile floor.
The walls are panelled with knot-free American pine.

View from the recessed garden room

Country home of Professor Albert Einstein. View into the large living room. Built-in cupboard in the background

Bathroom with tile floor and composition board on the walls

Country home of Professor Albert Einstein. View from the large living room to the open terrace. The three-panelled glass door can be completely opened.

The massive beam structure upon which the three upper storeys rest can be clearly seen.

Tourist Hotel for the Bavarian Zugspitze Railway, on the face of the Zugspitze, built in a record 35 days

Overall view of the construction

Tourist Hotel on the Zugspitze. View of the main central section with large beam supports. Designed by the Engineering Office of the AEG

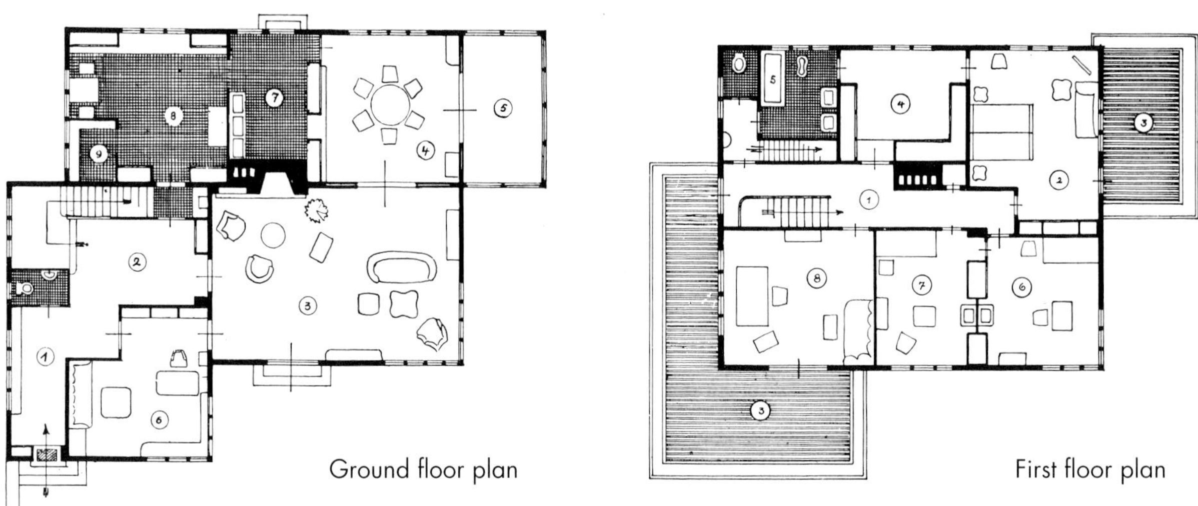

Villa, Am Rupenhorn, Berlin. View from the garden. Professor Karl Bertsch, architect, Berlin

Villa Am Rupenhorn. View of the entrance hall. Tile floors and seamless plywood panels

Living room corner with large windows

The open fireplace in the living room. The walls are panelled with seamless plywood.

Living room with built-in cupboard

Villa Am Rupenhorn. The large kitchen with built-in shelving. Tiled floor and walls

Casing on the wings

Sanitarium near Kassel. View during construction. The ground floor has been partially cased. The framing of the upper floor is clearly visible.

General view of the building

Sanitarium near Kassel. Rafters are in place

Earth Sciences Institute in Ratibor. Main view. Konrad Wachsmann, architect

Earth Sciences Institute in Ratibor. View of the north side. Extensive terracing on the building is provided for the placement of scientific instruments.

View of the laboratory

Earth Sciences Institute in Ratibor. The north side with its extensive laboratory. Staff apartments for senior staff are located on the upper storey.

View of the staircase

Earth Sciences Institute in Ratibor. Rear view showing entrances. This building is the first of a large complex.

Large dining hall with sliding windows

Beach Hotel on the Hengsteysee, Ruhr. View from the beach. City Building Councillor (retired) Hans Strobel, architect, Dortmund

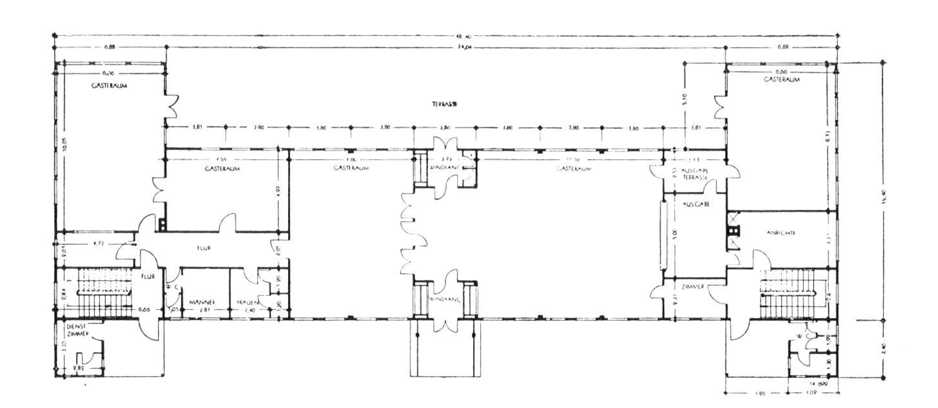

The main entrance

Beach Hotel on the Hengsteysee. Rear view showing entrances. The building is bevel-cased

A small hunting lodge near Dresden

A house in a worker's colony in the Dresden heath. Eugen Schwemmle, architect, Dresden

General view

A home near Heidenheim. The high stone foundation is necessary because of the steeply graded site. Eugen Schwemmle, architect, Dresden

View of the entrance

Kindergarten for 60 children. View of the play area. The main classroom is in the interior portion of the building and is illuminated with skylights. Heim and Kempter, architects, Breslau

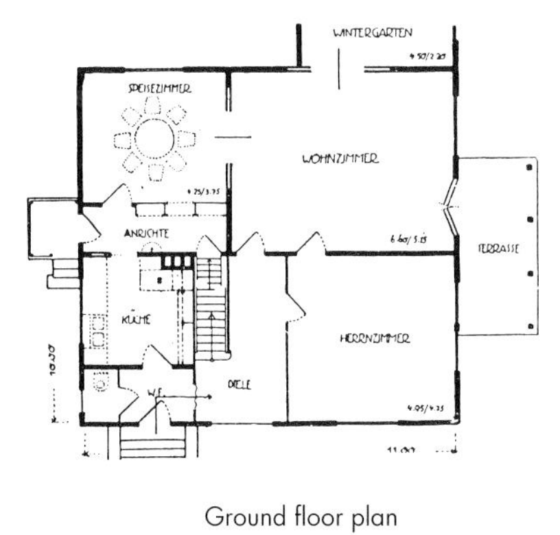

Ground floor plan

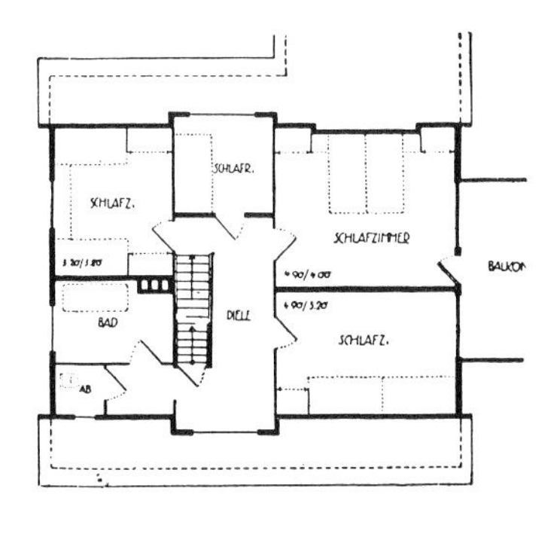

First floor plan

A house for a large family. Eugen Schwemmle, architect, Dresden

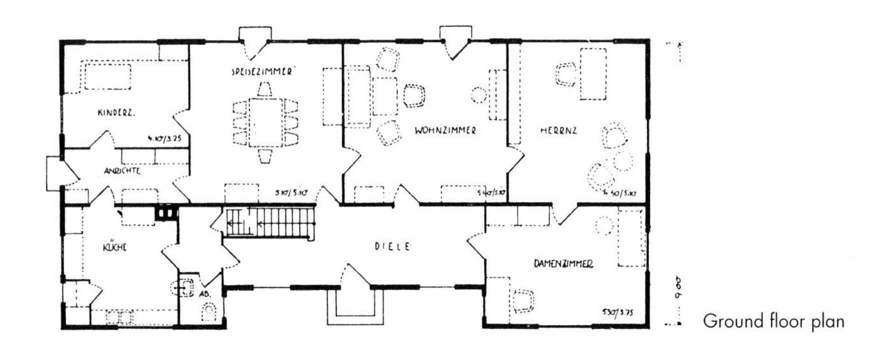

Ground floor plan

A house designed by Professor Hans Poelzig, Berlin. The outer wall is planked.

Detail of the frame

A plaster and wood frame house designed by Professor Paul Schmitthenner.
View of the entrance side

The framework is not very different from that of old wood frame houses.

House designed by Professor Schmitthenner using light-weight stone filler

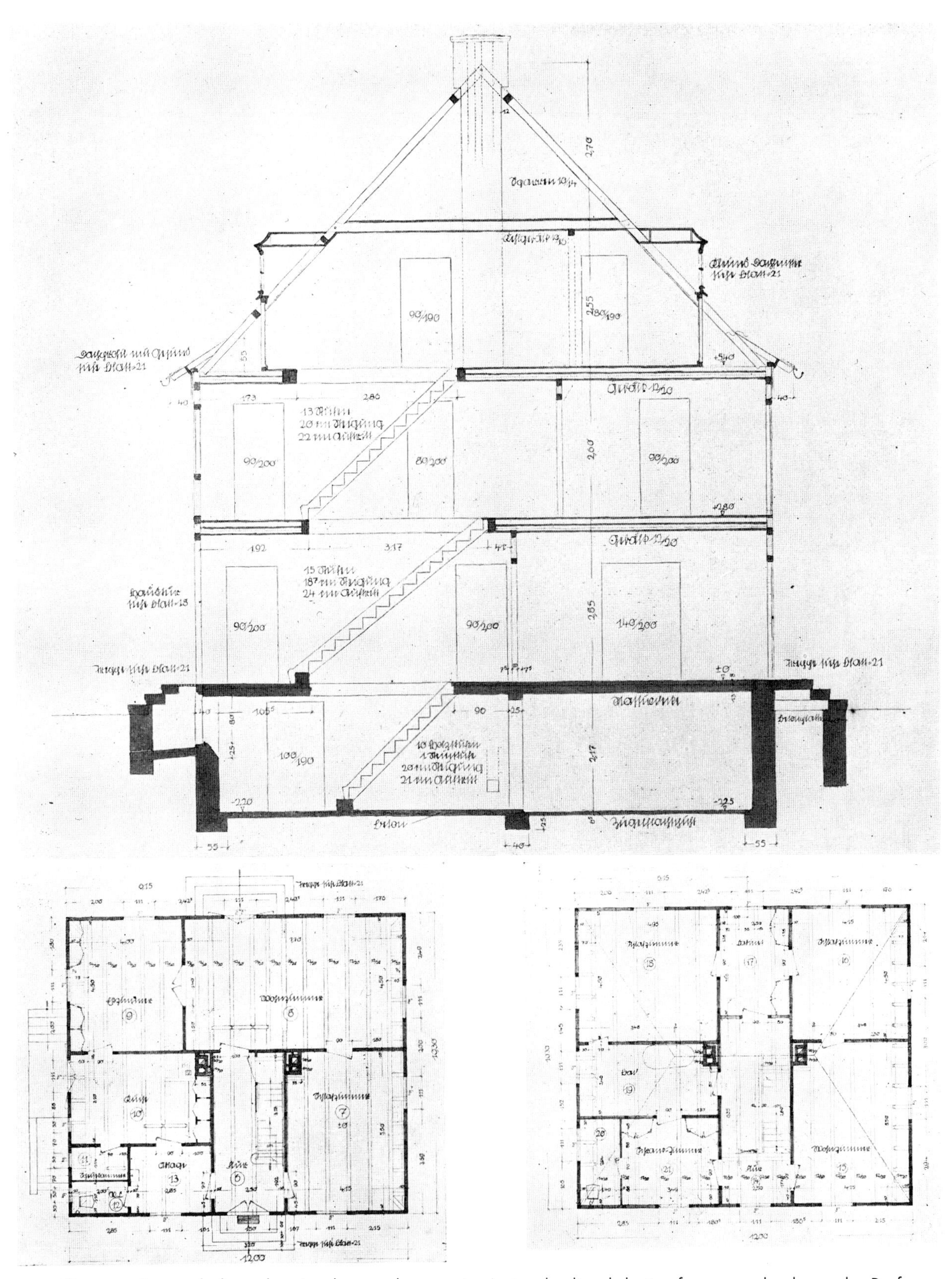

Cross-section and plans showing beam placement. A standardized design for a wooden house by Professor Schmitthenner, Stuttgart

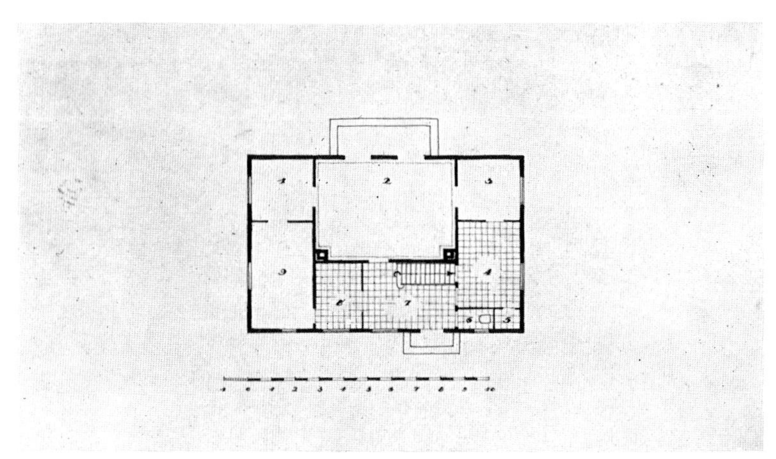

Ground floor plan

Home of the author Annette Kolb. Wood frame with plaster. Professor Paul Schmitthenner, architect

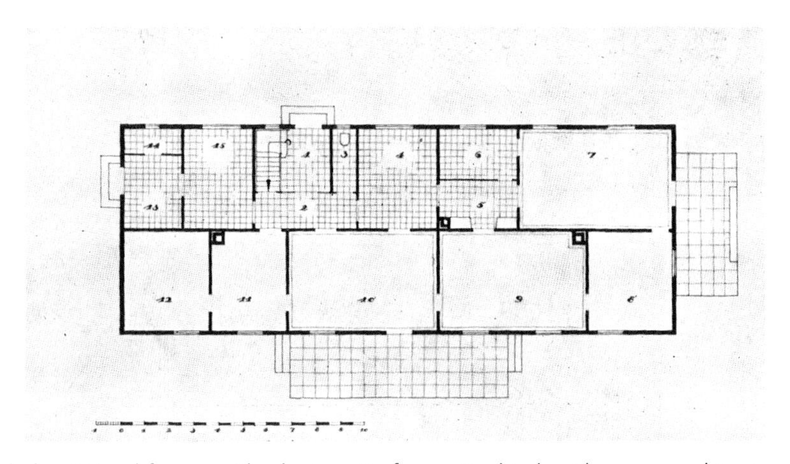

Ground floor plan

Home of the author René Schickele. Wood frame with plaster. Professor Paul Schmitthenner, architect, Stuttgart

The method of attaching the trusses to a central steel collar can be clearly seen. Construction design by Dipl.-Ing. Seiler, Leipzig

Dance pavilion in the Leipzig city park. View during construction, which was completed in 4 days. Professor Karl Bertsch, architect, assisted by Ditterle, architect

Interior and exterior walls are panelled with plywood.

Dance pavilion in Leipzig's city park. Partial view of the terrace

Detail of the cased trusswork.
Gymnasium in Saxony

A gymnasium used for exhibition purposes. City Building Councillor Ernst May, architect

Interior view of the gymnasium showing the stage

Gymnasium in Hettingen. Hans Zimmermann, architect, Stuttgart

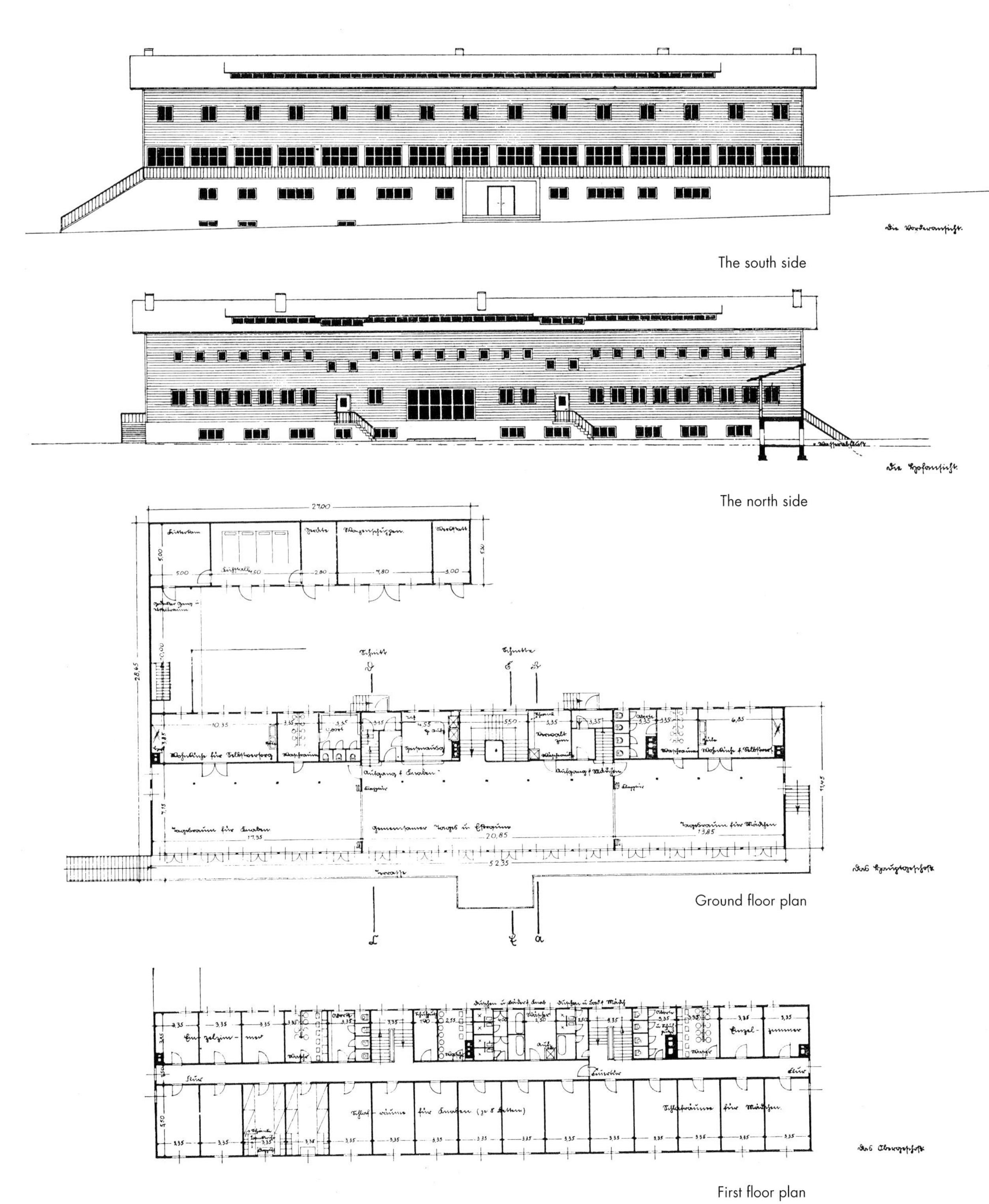

The south side

The north side

Ground floor plan

First floor plan

Design entered in competition for a youth hostel in the Riesengebirge. Konrad Wachsmann, architect

THE PANEL METHOD

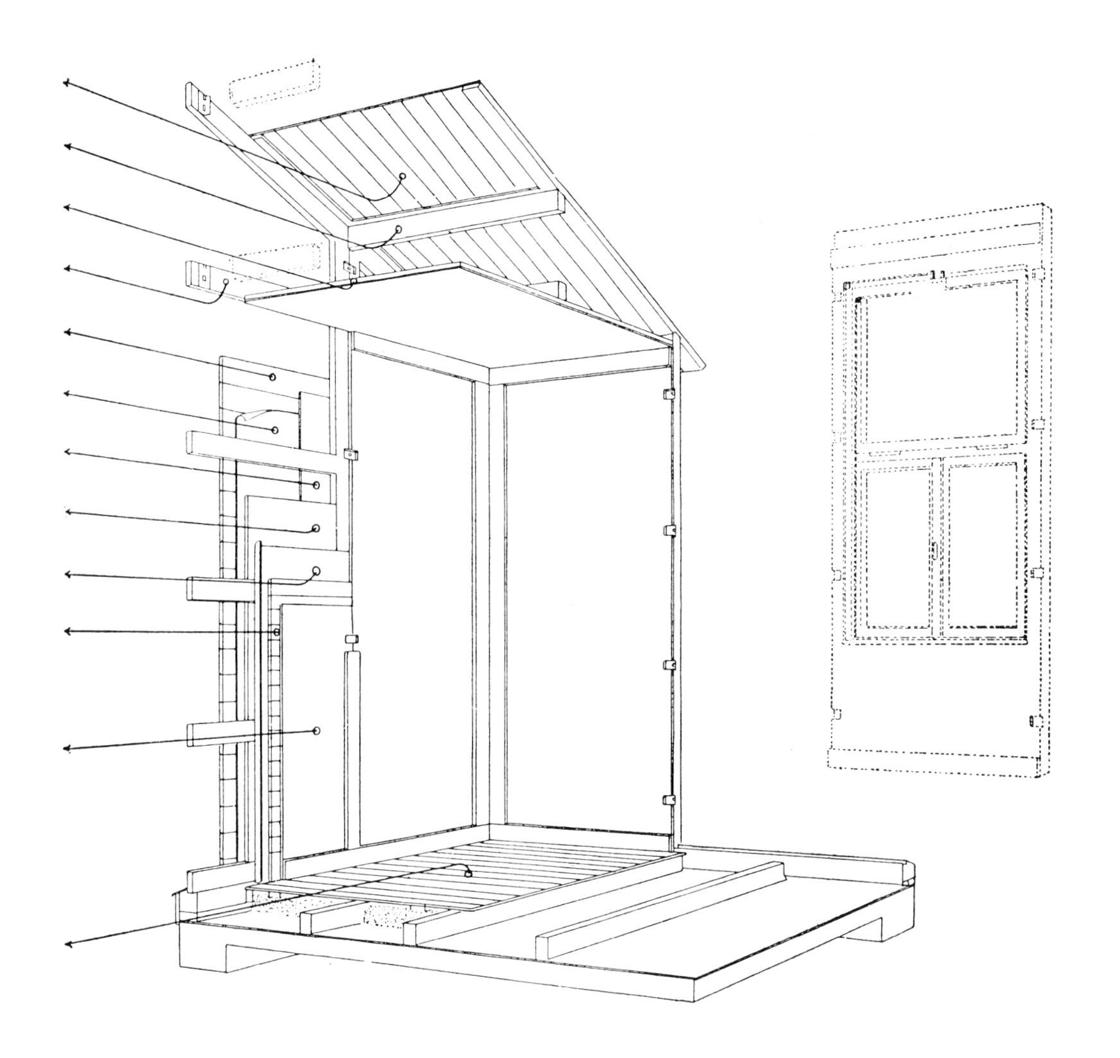

General view

Restaurant at the theater exhibition in Magdeburg. Professor Albinmüller, architect

The windows of the work room

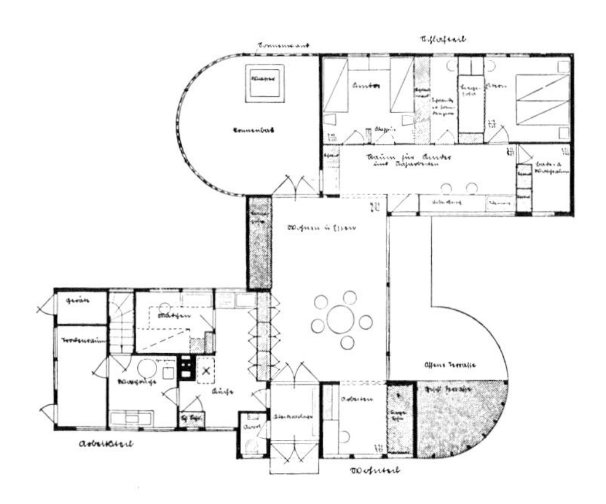

A single-storey house built with standardized walls and a flat, composition roof. Professor Hans Scharoun, architect, Breslau

The covered portion of the terrace

Tiled interior court showing the large, sliding windows of the main living room.

General view from the park

Large country home in Holland designed by Professor Henry van de Velde. View of the entrance. The roof is thatched with straw.

View of the main entrance

Single-family house built using the panel method. Garden side with covered patio. Professor Bruno Paul, architect, Berlin

Rehearsal hall of the ballet company of the Berlin Opera. View of the main entrance. Gésa Loerincz and H.H. Zschimmer, architects, Berlin

The covered entrance of the rehearsal hall

The main entrance

Rear view showing the stage of the rehearsal hall

View of the wide stage stairs

Side view. The rhythm of the evenly spaced beams is accentuated through color.

Office building, Berlin Public Transport Authority. Panel joints without stripping. Konrad Wachsmann, architect

Rear view showing the staircase which is provided with external illumination.

Office building of the Berlin Public Transport Authority, showing moveable glass partitions.

Model

Journalists' Union youth hostel. Silesia. Peter Paul Friedrich, architect, Berlin

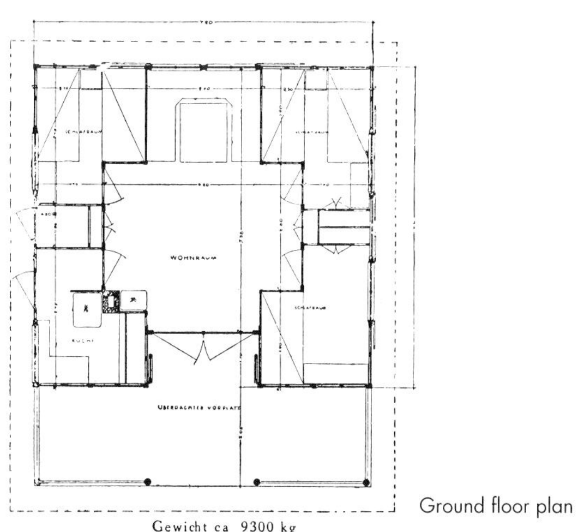

Gewicht ca 9300 kg

Ground floor plan

The living room alcove

Summer home for a large family showing the large covered terrace. Professor Hans Poelzig, architect, Berlin

The support element of the porch covering is a pegged beam in the roof.

Tennis court pavilion built for the former president of the German Railway System, Berlin. The covered terrace has a span of 9 meters. Konrad Wachsmann, architect

The built-in kitchenette

View from the large living room over the terrace. All four panels of the glass doors are open.

Partial view of a classroom wing with skylights that provide uniform lighting to each classroom.

Model school layout at a public health exhibition in Dresden

A multi-classroom school complex in Königsberg, Prussia. It consists of standardized buildings with pre-fabricated elements.

Above: A pre-fabricated classroom wing. Middle: Classrooms built in the 1880s.
Lower: Special purpose classrooms built for a horticultural school in Berlin.

THE LOG HOUSE METHOD

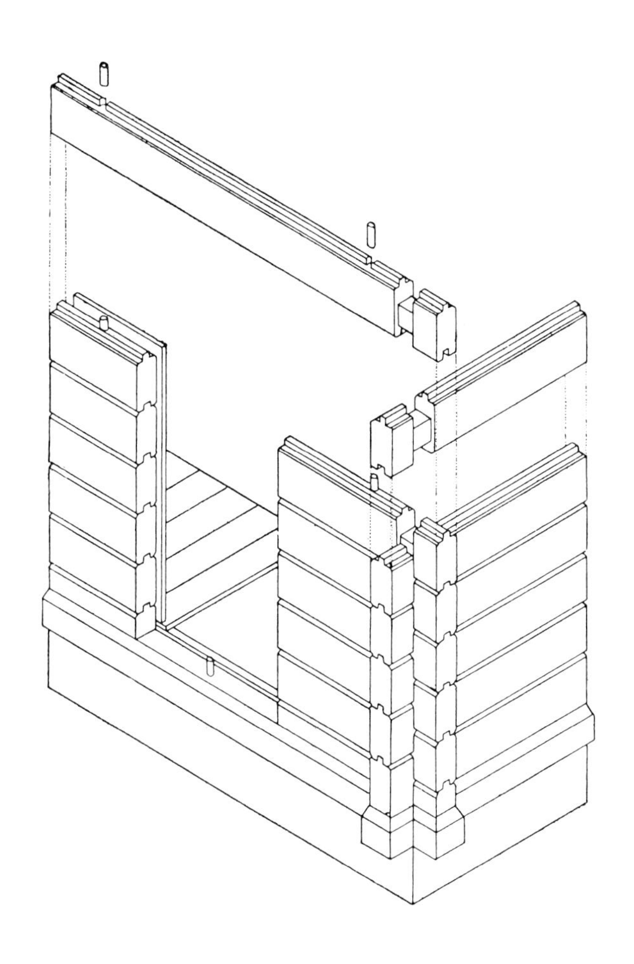

View of the wall framing

Director's residence in Niesky, Saxony. Garden side showing the 10-meter-long sun terrace. The ground floor windows are fitted with sliding shutters. Konrad Wachsmann, architect

Wall framing without roof

Director's residence in Niesky, Saxony. General view from the garden

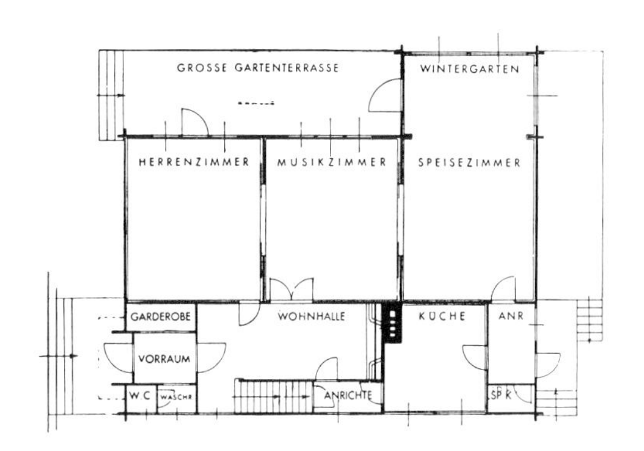

Ground floor plan. Rafters being raised

Director's residence in Niesky, Saxony. General view from the street

The large hallway showing the
6-meter-long window

Director's residence in Niesky, Saxony. View from the street showing the large hallway window

Upper storey window with one tilt window and two side-hinged windows

Director's residence in Niesky, Saxony. Main entrance with a projecting roof

Director's residence in Niesky, Saxony. Entrance hall with main staircase. The chimney forms a free-standing column. The wall panelling is visible.

Director's residence in Niesky, Saxony. Built-in shelving with sliding glass doors. Wall panelling with large plywood panels

Built-in sinks in the kitchen

Single-family house in Stuttgart. The extended eaves of the upper storey were built using the on-site method.
Hans Zimmermann, architect, Stuttgart

Single-family house in Stuttgart. General view from the street. Hans Zimmermann, architect, Stuttgart

Two-family home housing development in Dresden-Prohlis. City Building Councillor Oertel, architect, Dresden

Two-family home housing development in Neudamm. Hans Nieter, architect, Niesky, Saxony

Art dealer's country home near Berlin

Single-storey house with canopy roof. Hadda, architect, Breslau